23987
/

D1626202

3 9002 00021938 3

Digital Basics for Cable Television Systems

ISBN 0-13-743915-6

Pearson
Education

Hewlett-Packard Professional Books

Digital Basics for Cable Television Systems

Jeffrey L. Thomas
and
Francis M. Edgington

http://www.hp.com/go/retailbooks

Prentice Hall PTR
Upper Saddle River
New Jersey 07458
http://www.phptr.com

LIBRARY

REGIONAL
TECHNICAL
COLLEGE, LIMERICK

class no: 621.388 THO
acc. no: 23987

Editorial/Production Supervision: *Joanne Anzalone*
Acquisitions Editor: *Jeff Pepper*
Cover Design Director: *Jerry Votta*
Cover Designer: Talar Agasyan
Manufacturing Manager: *Alexis R. Heydt*
Marketing Manager: *Miles Williams*
Editorial Assistant: *Linda Ramagnano*
Technical Reviewers: *Even Kristoffersen, Jack Moran*
Manager, Hewlett-Packard Press: *Patricia Pekary*
Editor, Hewlett-Packard Press: *Susan Wright*

© 1999 by Hewlett-Packard Company

Published by Prentice Hall PTR
Prentice-Hall, Inc.
A Simon & Schuster Company
Upper Saddle River, New Jersey 07458

Prentice Hall books are widely used by corporations and government agencies for training, market-ing, and resale. The publisher offers discounts on this book when ordered in bulk quantities.
For more information, contact
Corporate Sales Department,
Prentice Hall PTR
One Lake Street
Upper Saddle River, NJ 07458
Phone: 800-382-3419; FAX: 201-236-7141
E-mail (Internet): corpsales@prenhall.com

All rights reserved. No part of this book may be reproduced, in any form or by any means, without permission in writing from the publisher.

All product names mentioned herein are the trademarks of their respective owners.

Printed in the United States of America
10 9 8 7 6 5 4 3 2 1

ISBN 0-13-743915-6

Prentice-Hall International (UK) Limited, *London*
Prentice-Hall of Australia Pty. Limited, *Sydney*
Prentice-Hall Canada Inc., *Toronto*
Prentice-Hall Hispanoamericana, S.A., *Mexico*
Prentice-Hall of India Private Limited, *New Delhi*
Prentice-Hall of Japan, Inc., *Tokyo*
Simon & Schuster Asia Pte. Ltd., *Singapore*
Editora Prentice-Hall do Brasil, Ltda., *Rio de Janeiro*

To Celeste for her love and support. J

Table of Contents

Preface

Digital technology is coming to a cable system near you. How will this new technology affect your day-to-day job? Why is it that you feel a bit lost when you open a cable television or broadcast trade magazine and find unfamiliar concepts, words, phrases, and acronyms? I certainly did when I started research for this book. Fortunately, I just asked my partner in this project, Francis Edgington. He has helped many of you with your digital signal cable measurement questions. We have learned an important lesson.

A system that keeps high analog performance standards may not provide the same reliability for digital signals. Worse, your analog proof-of-performance measurements do not help you troubleshoot or fix a digital signal problem. Digital video signals, in and of themselves, do not show you their contents or their quality. They recover from stress automatically, to provide the best picture, sound, or data to your subscriber, but may be close to crashing without your knowing it.

Our first goal is to acquaint you with the fundamentals of digital technology, system integration, and quality parameters as they relate to the delivery of **digital video over cable and optical fiber systems** so you can study what you need, when you need it. Our second goal is to leverage your basic knowledge into making realistic measurements on digital signals in your cable system with the tools at hand. The third goal is to help you recognize the symptoms of trouble caused by digital signals in your analog environment so you can efficiently get it operating again.

This book is for cable, telephony, and radio frequency (RF) and optical fiber system technicians and engineers who install, upgrade, and maintain analog- and digital-signal systems. To get the most out of this book you should be experienced or at least familiar with the signal technologies and testing of NTSC or PAL cable television systems. This includes some fundamental test instrument knowledge, such as the way a television channel looks in the time and frequency domains, but not necessarily the modulation theory behind the video formats. You should also understand the fundamentals of distortion and carrier-to-noise ratio in systems with multiple channels. You can pick up most of this background from the book *Cable Television Proof-of-Performance*, listed in the bibliography of the first chapter.

The chapters have been kept small to allow reading in one or two brief sessions. The beginning of each chapter outlines what you will learn. As a review, each chapter includes a summary and questions for review. The answers to the review questions are in an appendix.

Here is the contents by chapter:

- Chapter 1 introduces the differences and similarities between analog and digital signals.
- Chapter 2 teaches you how information is sent on carriers using modulation.
- Chapter 3 introduces the transformation of an analog signal into a digital bit stream.
- Chapter 4 discusses the distribution of signals, and how digital signals are layered to protect themselves from transmission harm.
- Chapter 5 shows how digital modulation is created and viewed.
- Chapter 6 digs deeper into the methods used for making a digital signal compact, resistant to transportation problems, and self-correcting.
- Chapter 7 summarizes the uses and attributes of several popular digital modulation formats and multiplexing schemes.
- Chapter 8 describes the measurement of digital signal quality.
- Chapter 9 helps you make digital signal power measurements.
- Chapter 10 helps you make burst power measurements.
- Chapter 11 shows how distortion, noise, and interference affect digital signals.
- Appendices contain a glossary of terms and acronyms used in the book, a performance and measurement map, a test equipment survey, answers to the chapter questions.

Many people helped to inspire and watch over our creation of this book. First is Helen Chen of Hewlett-Packard, who, along with her design team, organized the cable television digital test scenario in articles first published in 1995. Her firsthand tutoring of the strengths and weaknesses of digital signals and their transport were fundamental to the creation of this book. Next we wish to acknowledge the people who put long hours in reading and critiquing the manuscript: Ian Johnston, Jack Moran, Bill Morgan, and Ian Wright. We would also like to thank Randy Goehler, Dan Kahn, Even Kristoffersen, and David Whitton for their notes, updates, and encouragement along the way.

We hope you enjoy the reading. We have tried to keep the tone light, and the explanations simple. Our goal is to give you enough background in digital technology to help you make fast and efficient measurements. Please do not hesitate to let us know if we have succeeded, or where we can improve the material. Here are our Internet addresses:

jeff_thomas@hp.com

francis_edgington@hp.com.

JEFFREY L. THOMAS

FRANCIS M. EDGINGTON

1

Digital Signal Measurement Guidelines

This chapter introduces guidelines to help you understand digital signals in an analog world.

What you will learn

- Define analog and digital in cable communications
- How are digital signals different from analog signals?
- How does the nature of a digital signal affect your ability to make measurements?

Defining analog and digital for communications

If you asked your friends to define the terms analog and digital you would have as many answers as people interviewed. Everyone's experience with these concepts is different. Digital is not just reserved for computers and digital audio and video. Digital is integrated into most of the products and services we use every day, especially in communications equipment.

What is Analog?	For the purposes of this book, an analog signal is one which contains information that is added to a carrier signal using a modulation technique. All signals, whether they contain analog or digital information, are transmitted within an analog medium, such as a coaxial cable, a fiber optic cable or through the open space.

In this book we use the terms digital and analog to describe the nature of the signal. All communications, whether analog or digital are ultimately transmitted from one physical location to another, using analog transmission medium such as a fiberoptics, a radio frequency (RF) cable, a telephone line, a terrestrial RF and microwave wireless link, or a satellite relay. Even computer network digital messages move between distant locations through an analog physical layer.

The emphasis in this book is on measuring digital signals in analog media.

We perceive the world as analog

Analog means information that changes as continuously varying values. When we look, feel, taste, or hear, we do it with analog senses and analog interpretation. Our senses receive information as continuous values. Only the finest differentiation breaks down an image as discrete values, for example, the rods and cones of our eye's retina. But our brains don't "see" it that way. The brain's interpretation of visual information is analog, blending this "dot" information into a contiguous picture.

Digital signals are all around us. Communications equipment as familiar as a FAX machine, or as remote as a microwave link relay station, uses a combination of analog and digital technologies. Table 1.1 shows some examples of common technologies and their signal characteristics.

Communication system or appliance	Signal characteristics
FAX	The scanned page of printed information is converted to a digital data stream by the FAX machine. The signal is sent on the telephone line.
CD - audio	Digital, as a series of dots on the disk surface read by a laser into a microprocessor. The coding AAD, ADD, and DDD chart the analog (A) and digital (D) history of the signal from recording, mixing and distribution, respectively. The CD player converts digital data stream to an analog audio signal which is output to the stereo system
Video Tape Recorder, VHS format	Analog video signal laid down as diagonal magnetic stripes on the tape. The player provides either a baseband or RF signal to the television.
Cellular telephone	Analog and/or digital signal processing, transmitted over the air as modulated RF.
Internet browser	Digital bit stream from a telephone modem, cable modem, or dedicated network line.

Table 1.1. Digital applications in everyday communications.

Similar uses of analog and digital formats are in your cable television system, shown in Table 1.2.

Cable television system components	Signal characteristics
Satellite feed to head end	Digital or analog microwave signals over the air on a microwave carrier
Return path pay-per-view message from set-top box	Digital information modulated onto a return carrier dedicated to the purpose
Fiber feed for importing remote TV channels	Analog and digital signals lightwave modulated onto an optical fiber
NTSC, PAL or SECAM television	Analog video modulated onto a carrier and transmitted over the air or on a cable or fiber
Cable modem (also on phone line)	Digital baseband modulated onto a burst carrier
Hybrid fiber coax (HFC) signal distribution	Analog and digital signals transmitted as analog signals

Table 1.2. Digital and analog use throughout cable television systems.

Analog is the non-quantized storage and transmission of data. **Digital** is the quantized storage and transmission of data. It is easier to distinguish between information sent as discrete values (digital) or continuous values (analog). Remember LP (for long play) records? The groove cut in the vinyl surface of an LP are analog. Close inspection with a microscope would show smooth curves in the groove's walls, as illustrated on the left side of Figure 1.1. The physical ripples are created from the sound waves themselves.

Figure 1.1. The groove of the vinyl record and pits of the CD.

Vinyl records became part of history when the audio compact disk, or CD, came to the market. The CD stores the music as digital information, placed as pits and spaces on the smooth surface of the CD. The signal, unlike the vinyl record, is not a physical representation of the sound waves, but a digitally coded signal, created by signal processing from the musical sounds. The pattern is read by the CD player as laser reflections, as shown on the right side of Figure 1.1, and reconstructed as digital information, called a digital **data stream**. The data stream is decoded into an analog audio signal.

The signal presents itself to your ear as a reproduction of the original analog audio information using analog audio components, such as preamplifiers, power amplifiers, equalizers, copper cables, and loud speakers.

Digital signal

A digital signal, defined for this book, is data stored and transmitted as a well-defined set of quantized values. Quantized means that the numbers have been saved as a collection of discrete digital values. Using the binary numbering scheme is just one way to digitize numbers.

Digital and analog, similarities and differences

Here are some general rules, or guidelines that distinguish digital and analog signals. Each is represented by an icon and phrase in the left-hand column of this text. Throughout the book, you will see these icons when a point is made that re-enforces the guideline. Some of these are common sense, but others need more explanation.

Transmission of analog & digital	**All signals, whether digital or analog are transmitted using analog techniques. Even the cable that connects your printer to your computer sends bit after bit of data as a varying voltage signal over an analog medium, copper wire.**

Distributed in analog

Digital and analog signals are transported in analog media.

A digital signal is sent down the same cable as an analog signal, separated only by frequency. The cable or hybrid fiber/cable system is an analog transmission path. It does not matter to the cable distribution system whether the signals are analog or digital; their treatment by the retransmission hardware is the same. The analog and digital signals survive corruption by keeping within their allotted channel frequency bandwidth and by maintaining power levels consistent with the analog cable system performance requirements.

All signals are in channels

Digital and analog signals ride on a signal carrier and are confined to a specific bandwidth.

To be transmitted, all signals, whether digital or analog, need a **carrier** signal. The carrier signal has no information of its own, but provides steady energy at a stable frequency for reliable transmission. Information is **modulated** onto a carrier allowing bandwidth for the modulated information. This is **channelization**. Figure 1.2 shows analog and digital signals residing side-by-side in channels in a system's frequency spectrum. The information of the analog or digital signal is added to this carrier using modulation. The frequency band occupied by the modulation, also known as **transmission bandwidth**, surrounds the carrier. The carrier and its modulation must stay within the confines of the allocated channel bandwidth.

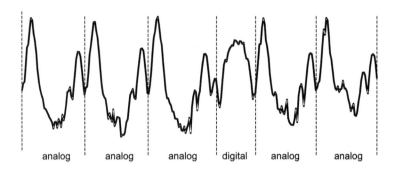

analog   analog   analog   digital   analog   analog

**Measurable
parameters**

**Figure 1.2. Analog and digital television channels in the frequency domain of a
cable television distribution system.**

Digital signals have unique measurement parameters.

The performance of a signal, that is, its specified quality, is quantified by measuring one or more parameters. The analog performance
parameters, such as C/N and CSO/CTB are familiar. But parameters
for digital modulation quality, such as MER, EVM, and margin, may
not be. The measurement parameters are divided into general application requirements:

- Analog performance (called proof-of-performance in North
  America)
- Digital modulation quality
- Data quality

In general, most of the cable television analog performance measurements apply to analog and digital signals alike. Meeting analog performance is a good starting point for digital signal quality, but many
analog tests are not useful for digital signals at all. Digital signal technologies bring a whole array of new measurement requirements to
cable television. **Bit error rate** (BER), **modulation error ratio**
(MER), and **error vector magnitude** (EVM) are a few measurements only applicable to digital signals. Discussion of these parameters starts in How an Analog Wave Becomes Digital Data, Chapter 3
beginning on page 33.

**Victims of
distortion, noise,
and interference**

Digital and analog signals produce and are the victims of distortion, noise, and interference. **Distortion** is an unwanted
change to a signal as it passes through a distribution system. The un-
wanted effects of distortion, noise and interference degrade the signal's quality. **Noise** is the random signal energy added to the signal

by active devices in its distribution which degrades the signal quality, usually by randomizing the signal's information so much it cannot be recovered. **Interference**, also called **ingress**, is noise and/or distortion which comes from sources outside the system such as over-the-air broadcast and ham radio transmissions.

Every signal is subject to the distortion produced by other signals or system hardware. In systems with many digital signals, distortion looks like noise because the digital signals causing the distortion are noise-like themselves. Every signal can produce distortion that interferes with other signals. Signals interfere with others by leaking power outside their allocated bandwidth. This type of distortion is called **out-of-channel**. When a signal creates an unwanted signal within its own bandwidth, the distortion is called **in-channel**.

Distortion, noise, and interference are discussed in Chapter 11 beginning on page 211.

Digital signal can send more information

For a given bandwidth, a digital signal can send more information than an analog signal. Six or more digital television channels can be transmitted in the same channel bandwidth now occupied by a single PAL or NTSC analog channel. This bandwidth economy requires digital data **compression**. Data compression, not to be confused with power compression, is reduction of signal data by eliminating repetitive and non-essential information.

Many services, besides television, use digital signal processing. Here are a few examples:

- Teletext
- Pay per view and video on demand
- Game subscription
- Internet access
- E-mail
- Telephony

Digital technology makes many of these services cost-effective. In fact, video motion picture transmission, as high definition, or compressed groups in a single channel, would not be possible without signal compression. This topic is discussed in , Chapter 7 beginning on page 99. **The emphasis in this book and these guidelines is on the delivery and maintenance of quality digital video transmission**. But many of the topics and measurement procedures also include aspects of digital information transmission that will help you understand the integration of computer networking into your systems.

Repair data at the receiver

Some digital signals carry information that allows damaged data to be repaired at the receiver.

When a signal is converted into digital bits, a certain percentage of the code is reserved for specially coded data that is used at the receiver for repair of the signal's information. The term commonly used for this process is error correction. See Error Correction, Equalization, and Compression, Chapter 7 beginning on page 99.

Digital signals crash without warning

Digital signals keep consistently high quality, but can crash with little warning.

Analog signals manage to convey information even if they are badly distorted. You can watch a television show through the snow of poor carrier-to-noise or the zig-zag lines of intermodulation distortion and still be entertained by the show. A digital signal maintains a higher level of quality than an analog signal when subjected to the same amount of transportation **impairment**, such as linear distortion, nonlinear distortion, noise, and interference. But at some level of impairment, long after customers begin to rebel against the poor analog picture quality, the digital signal crashes. This is shown graphically in Figure 1.3. The middle region represents the superior robustness of digital video over analog signals especially in the presence of noise and distortion. In video this crash is a black, scrambled, or frozen picture, sometimes with a buzzing sound. In data transmission, a crash means an application cannot run, or a file cannot be opened.

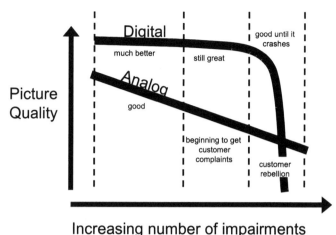

Picture Quality

Increasing number of impairments

Figure 1.3. Analog and digital signal response to increasing impairments.

This is called the **cliff**, or **waterfall** effect. A digital signal has the capacity for self repair in real-time. The technologies that make it possible for a digital signal to hold out against adversity are

- Digital signal processing
- Signal compression
- Error correction
- Adaptive equalization

Digital signal processing is covered in How an Analog Wave Becomes Digital Data, Chapter 3 beginning on page 33, and Viewing Digital Modulation, Chapter 6 beginning on page 77. Compression, error correction, and adaptive equalization are covered in Chapter 7 beginning on page 99.

Digital information is hidden

Digital signal content is hidden deep in the signal. You are used to tapping into your cable system almost anywhere and looking at the familiar television channel spectrum or picture on your TV test monitor. Not so with the digital channel. To be robust and compact, the digital signal is "homogenized" until it is more like noise than an analog-modulated signal. Compression, and error correction help conceal the signal. This signal processing is called digital **encoding**. The modulation that puts the digital baseband information onto the carrier does so by spreading the signal almost uniformly over the entire bandwidth of the channel, further disguising the nature of the signal. See Chapter 7 beginning on page 99, Chapter 2 beginning on page 15, and Chapter 6 beginning on page 77.

Multiplexing combines several signals into one

Multiplexing combines several signals into one. Multiplexing means combining more than one signal in the same transportation medium, usually for more efficient use of the available frequency spectrum. A cable television combiner, which adds channels to the trunk cable, multiplexes signals using **frequency division multiplex, or FDM**. FDM is just a fancy way to describe the function of a broadband RF cable distribution scheme as shown in Figure 1.4.

Digital information is also multiplexed by combining baseband data streams together into a single data stream. The zipper symbol in the left margin icon is a reminder of how data streams can be combined. These streams may, in turn, be combined in a broadband transmission medium such as wireless or cable, using any one of many multiplexing techniques. These techniques can usually be recognized by the word "division" in their name. Examples are **time division multiple access, or TDMA,** and **code division multiple access,** or **CDMA**.

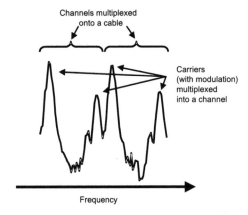

Channels multiplexed onto a cable

Carriers (with modulation) multiplexed into a channel

Frequency

Figure 1.4. Analog channels in a cable system.

Multiplexing is covered in Distributing, Layering, and Multiplexing, Chapter 4 beginning on page 53.

Performance and measurements preview

Measurements are necessary to judge system performance. In an analog signal cable television system, analog performance is the primary set of measurements and performance levels required. When digital signals are added, a new set of interactive performance gauges and measurement tools are introduced. Two ways are used to look at performance and measurements: Where to measure, and what to measure. The where to measure is shown in Figure 1.5.

The signal path through the forward direction is shown for a typical hybrid fiber/cable system. The only digital addition is the digital-to-analog blocks near each end of the system. Power and spectrum measurements, many of which are made in the analog performance suite of tests, are made in this RF portion of the system.

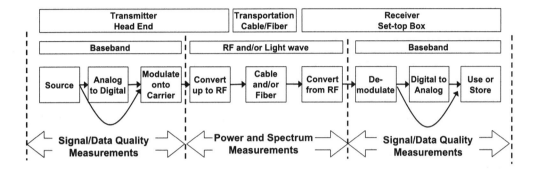

Figure 1.5. Typical signal transportation through a system.

Another way to look at measurements that helps you see the relationships between analog and digital signals in the RF domain, is shown in Figure 1.6. This is a sample performance and measurement map of the measurement parameters. (The full performance and measurement map is provided in Appendix B, beginning on page 247.) Every block is either a measured parameter, such as CSO/CTB distortion, noted by the dotted boxes, or a system performance characteristic, which may or may not be a measured parameter. Examples of performance characteristics are gain, noise, and interference.

The lines and positions of the blocks show their relationship. Blocks connected from the left have a direct influence on the right-hand block. As an example, look at carrier-to-noise to the left of center in the figure. The analog carrier level and the system noise determine C/N, so they are shown connected on the left.

Boxes connected directly to a performance, such as the way Proof-of-Performance is attached to Analog Channel Quality, indicate that the quality of the parameter is measured by that parameter. Another example, is the Digital Channel and Data Quality, where bit error rate (BER) is a direct measure of the signal quality, and the cliff effect is shown to be an indicator of quality.

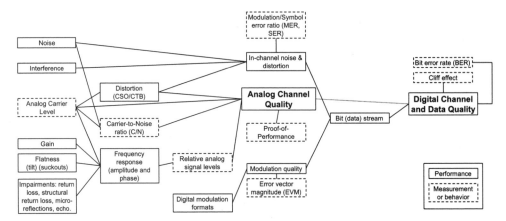

Figure 1.6. Performance and measurement flow in a cable television system.

The rest of this book uses subsets of the full map in the appendix to illustrate performance and measurement basics. Measurement instrumentation is added to these maps in the measurement chapters.

Every map reemphasizes a key point: the system performance that provides high analog signal quality is a mandatory starting point for your system to provide superior digital signal quality.

Summary

This chapter introduced some guidelines for helping to understand the behavior of digital signals, especially in relation to the familiar cable television analog signals. These guidelines are better defined and their messages reinforced throughout the remainder of the book. The left margin logos help you spot guideline material.

It is almost as important to know where to make the measurement as it is to know what measurement to make. A list of the measurements covered in this book show the suggested test equipment and the location of the tests.

Digital and analog signals ride on a signal carrier and are confined to a specific bandwidth.

 Digital signals have unique measurement parameters.

 Digital and analog signals are transported in analog media.

 Digital and analog signals produce and are the victims of distortion, noise, and interference.

 For a given bandwidth, a digital signal can send more information than an analog signal.

 Some digital signals carry information that allows damaged data to be repaired at the receiver.

 Digital signal content is hidden deep in the signal.

 Digital signals keep consistently high quality, but can crash with little warning.

 Multiplexing combines several signals into one.

Questions for Review

1. What are the characteristics of an analog signal (for the purposes of this book)? (check all that apply)
   a) continuously varying amplitude over frequency
   b) continuously varying amplitude over time
   c) signal may be discrete amplitude values over time
   d) the energy of the signal is spread over frequency

2. What characterizes a digital signal? (check all that apply)
   a) the amplitude of the signal is limited to a set of specific voltages
   b) the amplitude levels of the signal can easily be measured by a spectrum analyzer
   c) the energy of the signal is spread over frequency

3. The guidelines outlined in this chapter are always true. True or False?

4. Generally, digital signals hold their quality better under adverse transmission conditions than analog signals. True or False?

5. Analog and digital signals are distributed through an analog media such as the coax cable and amplifiers of a cable television distribution system. Which of the following systems are sending digital signals? Check all that apply.
   d) Computer sending print information to a local printer
   e) Computer sending print information to a network printer in a remote site.
   f) Hybrid fiber-coax head end to node transport in a cable television distribution system.
   g) The audio cable output of a compact disk player.
   h) The FAX output to a telephone line.

6. It is easy to tell what program material is being sent on a digital signal. True or False?

Selected bibliography

1. Dana Cervenka, "Designers Pour "Smarts" into Digital Test Gear," *Communications Engineering & Design*, Chilton Publications, New York, October 1996.
2. Francis M. Edgington, "Preparing for In-service Video Measurements," *Communications Engineering & Design*, June 1994.
3. William Grant, *Broadband Communications*, GWH Associates, 1996.
4. Nicholas Negroponte, *Being Digital*, ISBN 0-679-43919-6, Random House, Inc., New York, 1993.
5. Andrew S. Tanenbaum, *Computer Networks*, 3rd Edition, ISBN 0-13-349945-6, Prentice-Hall, Inc., 1996.
6. Jeffrey L. Thomas, *Cable Television Proof-of-Performance; A Practical Guide to Cable TV Compliance Measurements Using a Spectrum Analyzer,* ISBN 3-13-306382-8, Hewlett-Packard Press, Prentice-Hall, Inc., 1995.

2

Modulation and Frequency Management

The purpose of an electrical signal is to convey information to a user or for storage. This chapter explains how the modulation of a carrier carries information.

What you will learn

- How does information get onto a signal?
- What is a carrier?
- What is modulation?
- How does a signal look in the time and frequency domains?
- Can analog cable television measurements be used on digital signals?

Preparing information for distribution

When analog information is created as an electrical signal, its first life is as a **baseband signal**. A baseband signal, such as a studio television camera signal or a local area network input, is usually not distributed over great distances. It has a limited frequency content, 10 MHz, for example, and is not combined with other signals on the same cable. When a signal is an RF signal, also called a **channel**, it is made up of two elements: a **carrier**, and its **modulation**. A **carrier** is a signal capable of being imprinted with information, usually in the form of modulation. The carrier itself does not contain information. **Modulation** is the process of putting the information contained in the baseband signal on the carrier.

Information

Information is sent when the recipient cannot predict the content. A signal carrier, by definition, sends no information, so it does not change in power level nor frequency over time. It is a steady-state, or continuous, signal.

Baseband signals are modulated onto different carriers, spaced in frequency to give each one room without interfering with its neighbor-

ing signals. These processes are called **channelization** and
multiplexing.

**Baseband
signal**

**A baseband signal is one whose frequencies are those
generated when the signal was first created. Usually the
highest frequency in a baseband signal is also its bandwidth.
The lowest is close to zero Hz, or DC. An example is the video
output of a television camera or video tape recorder. Its
highest frequency, therefore its bandwidth, is either 4 or 6
MHz, depending upon which TV format is used.**

The carrier concept

**Carriers carry
information**

For some reason only nature and physics can explain, the sine wave
signal is an efficient carrier of energy from one point to another,
whether it be over the air, or on a fiber optic or electrical cable. Its
power level is at a constant average, and is confined to a single frequency. To see how a sine wave is generated, look at Figure 2.1. A
rotating stick has its shadow projected on the observer's screen. The
shadow of the stick moves up and down, getting longer and shorter
as the stick rotates. The shadow changes direction from the middle as
the stick dips below its pivot point. The motion of the tip can be plotted over time.

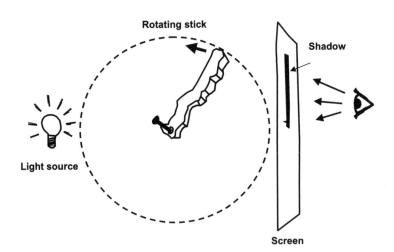

Figure 2.1. A rotating stick and its shadow.

This is illustrated in Figure 2.2. The amplitude of the shadow of the
stick is plotted along the right-hand side as a function of time, as if

the screen were moving to the left. The tip represents the amplitude of a signal any instant in time. Point 1 is the starting amplitude. Point 2 is casting a minimal shadow, at zero amplitude, so it is called the zero crossover point. Point 3 represents the lowest amplitude swing. This represents the greatest magnitude of the stick's swing as shown on the corresponding position of the time plot. Each full rotation of the stick has two zero crossings, and two points where the maximum magnitude is reached. A full rotation is called a single cycle. If the stick rotates at the same speed, called a constant angular velocity, each cycle takes the same amount of time.

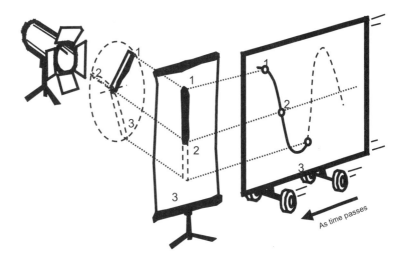

Figure 2.2. The rotation of the stick as a function of time.

Now apply this concept to the representation of an electric signal. Rather than a stick, think of a rotating line, or **vector**, the length of which is proportional to the signal voltage. The vector is anchored at 0 volts at one end called the **origin**. The rotating end of the vector is usually represented by an arrow head. The distance between the anchor point and the arrow head is called the vector **magnitude**. These terms are illustrated in Figure 2.3 on page 18. The rotating vector drawing on the left of the figure is called a **vector diagram**. The rotating vector, by convention, starts at zero degrees at a position corresponding to three o'clock. The vector rotates counter-clockwise, sweeping through 360 degrees for a single cycle. The vertical motion of the vector's arrow tip plots a sine wave as it rotates just like the shadow in Figure 2.1. The plot to the right shows magnitude versus the angle of rotation, measured in degrees or radians. The vector magnitude corresponds to the maximum swing up and down, points 3 and 6. Point 1 and 4 are the zero crossings. Points 2, 5, and 7 illustrate random positions of the vector and how they map into the sine wave.

Degrees and radians

Given a compass, you can easily measure the angle of rotation of a vector in degrees. But in the graphical and mathematical discussion of signal vectors, it is convenient from a mathematical viewpoint to use the radian. How do degrees and radians relate? There are, by definition, 2π radians in 360 degrees. The Greek letter π, or pi, represents a constant roughly 3.1416 dimensionless units. In discussions of rotating vectors it is no small coincidence that 2π times a circle's radius equals its circumference.

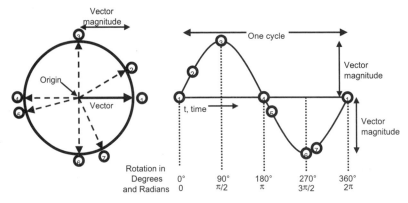

Figure 2.3. Definition of rotating vector terms.

A signal composed of a single sine wave is often called a **continuous wave** signal, or **CW** signal. A CW signal

Continuous wave has no bandwidth

- is not modulated, therefore does not carry information
- has a constant average power level
- has a constant frequency with a sine wave shape
- puts no energy in any frequency but its own

These characteristics make a CW signal perfect for use as a "carrier" of information.

Practical uses for CW carriers

The earliest electronic communications used a noisy CW signal in an on/off mode to send Morse code over the air or over wire to a receiver where a human interpreted the code. Although the carrier was CW the technique for sending code was early amplitude modulation. Not too many years ago cable television used CW carriers as pilots stationed at various places in the broadband cable spectrum. These pilots were usually without modulation so that their amplitude can be monitored quickly and accurately by circuits that automatically adjust the gain and slope of the trunk and feeder amplifiers for changes in forward path flatness and local changes in ambient temperature. (Now this work is done by monitoring modulated signals; optimizing the spectrum efficiency of the system.) The pilots traditionally did not convey information, but the action of the distribution system on them makes changes to their amplitudes, which can be measured and acted upon. Today pilots can be modulated, and the system uses its peak as the guiding level.

Modulation

Modulation is the process of adding information to a carrier. The information is usually in the form of a baseband analog or digital signal. These baseband signals are added to the carrier in such a way that the receiver of the modulated signal can separate the baseband signal from its carrier to access the information. Modulation acts on the carrier's amplitude, frequency, or phase characteristics one at a time or in combinations.

Modulation	**Modulation is the process of changing a carrier signal's frequency, amplitude, or phase, or some combination of these characteristics, directed by a baseband signal. The baseband signal contains the information message to be transmitted.**

The equation in Figure 2.3 shows how. The equation is explained in "Why use equations?" on page 20.

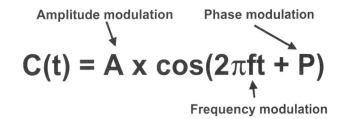

$$C(t) = A \times \cos(2\pi ft + P)$$

Amplitude modulation

Phase modulation

Frequency modulation

Figure 2.4. Modulation types.

Why use equations?

To understand the modulation parameters and measurements in the coming chapters, it is necessary to use the correct terminology and expressions. One of the most important concepts is the mathematical term "function." When a parameter, such as voltage, changes over time, it is said to be a function of time. This is written as $V(t)$. Do not be put off by the jargon. $V(t)$ means that the value, voltage, V, is changing in some way over time, t. If $V(t) = 1$ says that the voltage stays constant at 1 no matter what time it is. $V(t) = \cos(2\pi ft)$ means that the voltage changes as a sine wave function over time.

To visualize the relationships between carriers and their modulation it is necessary to introduce some basic trigonometric formula and standard engineering notation. Many measurement techniques and parameters are based on an understanding of these principles. The carrier's sine wave in Figure 2.3 is expressed as the trigonometric function $\cos(2\pi ft)$. This expression says that the carrier, $C(t)$, is function of time, t, in seconds, and equals the cosine of a constant, $2\pi f$, where f is the frequency in Hz, times time, t. The cosine function generates a sine wave just like the sine function, but starts out at 1, not 0, when the argument $2\pi ft$ is zero. It is easier to plot.

The $2\pi ft$ says that the cosine changes over time, t, in seconds at an angular velocity, $2\pi f$, in radians per second. The cosine function requires radians to be calculated. The radian unit is discussed on page 18.

Modulation of the carrier occurs when any one or combination of the factors A, P, and/or $2\pi f$ are also changing over time. If A, P, or f change, their change affects the carrier's wave, each in its own way. Any unique and well-defined modulation technique is referred to as a modulation **format. Amplitude modulation** (AM) occurs when the value of A changes, **frequency modulation** (FM) when f changes, and **phase modulation** (PM) when P changes in the equation of Figure 2.4. A, P, and f can be made independent functions of time. These are designated modulation signals A(t), P(t), and f(t), respectively. It is important to see the effect on the carrier as a voltage changing over time by each of these parameters to understand how digital modulation appears in both the time and frequency domains.

Modulation format

Format means an arrangement or plan. A modulation format is a specific technique for adding information to a carrier. The format name usually gives a hint on the method used, such as in "amplitude modulation."

Amplitude modulation

If a carrier is amplitude modulated, the energy of the signal changes with the modulation. The rotating stick in Figure 2.2 would change length as it rotated. The time domain waveforms in Figure 2.5 illustrate this. In (a) the carrier is shown as a CW signal. On average the power over time is not changing. The modulation baseband waveform in the specific case in (b) is another sine wave, although it could be any shape, periodic or random. The amplitude modulation process produces the signal in (c) with the amplitude of the signal varying in strength as a function of the modulating waveform. The average power of the carrier is changed as a function of the modulation base-band signal. If the modulation waveform is a non-periodic function such as voice, code, or music, the AM signal changes even more radically in power level. In fact, the best and most radical illustration of power changes with AM is the transmission of Morse code before the days of carrier modulators; the carrier was simply turned on long times for dashes, and brief times for dots.

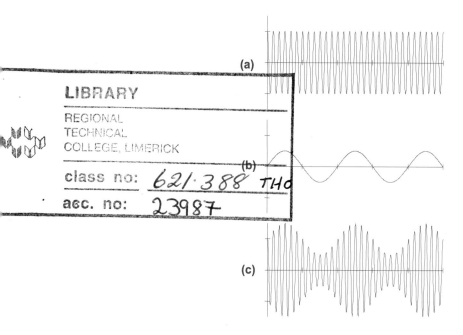

LIBRARY

REGIONAL
TECHNICAL
COLLEGE, LIMERICK

class no: *621·388* THd

acc. no: *23987*

Figure 2.5. Time display of carrier amplitude modulation.

Another common example of AM is the video carrier in the PAL- or NTSC-formatted video signal. In these waveforms, the amplitude modulation actually carves out the frame and line information from the maximum carrier level towards the zero voltage level. This is illustrated in Figure 2.6. The envelope of the video modulation is seen on top and bottom; the unmodulated carrier peaks are represented by the dotted line at the modulation peaks. AM, used in commercial radio transmission, varies above and below the nominal carrier level as in Figure 2.5. Television video information is modulated onto the signal by actually reducing the carrier signal level. This form of amplitude modulation in analog television video allows simpler design of the power amplifiers used for broadcast and distribution, and more important to the popularity of TV's low-cost, mass-produced television receivers.

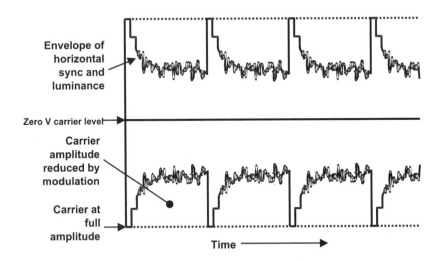

Figure 2.6. The AM waveform of a PAL or NTSC television video carrier.

Frequency and phase modulation

The sound, or aural, signal in the composite television channel is a frequency modulated signal. Although this signal is within the television channel bandwidth, the aural carrier is a separate signal **combined**, or added, to the video carrier before transmission. Commercial "FM" radio is frequency modulation. Both phase and frequency modulation change the angle of the carrier vector as it spins at the carrier frequency.

In the example of plotting the tip of the rotating stick, the amplitude does not change with FM and PM angular modulation. The stick changes speed as it rotates, having the effect of wobbling side-to-side from the viewpoint of the carrier. The shadow of the tip is always at the carrier level.

Because amplitude modulation changes the signal power, and angular modulation changes the frequency, there is little interference, or crosstalk, between them. This signal separation is an important concept to remember in the discussion of techniques for digital modulation in Viewing Digital Modulation, Chapter 6 beginning on page 77.

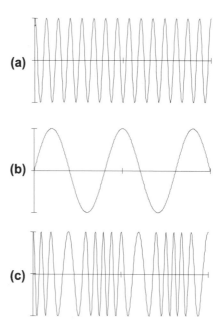

Figure 2.7. The effect of angular modulation on a carrier when the modulation signal is a sine wave.

The difference between frequency and phase modulation are the circuits used to convert the modulation signal into the angular changes. Their time domain effects are all but indistinguishable when the modulation signal is a smoothly varying signal such as a sine wave, as illustrated in Figure 2.7. The carrier is a CW signal in (a). The modulating signal, (b), affects the carrier in (c) by smoothly changing its frequency, increasing frequency on the upward slope of the modulation signal, and decreasing the frequency on the downward slope. But when dealing with sharp, impulsive baseband modulation as in digital signals, the results are quite different, as the following figure shows.

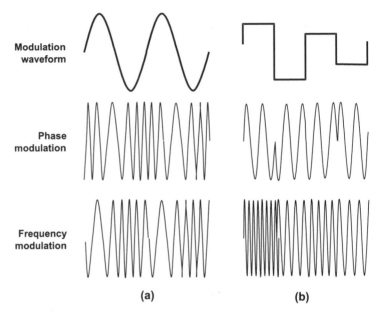

Figure 2.8. Time domain responses to different types of modulation signals.

Sharp changes in a signal's waveform cause signal bandwidth to widen

Figure 2.8 shows how phase and frequency modulation responds to sine and square wave modulation signals. The modulation waveforms in the signals of the (a) column respond with smooth changes to the modulation signal. The waveforms in column (b) are characteristic of the sharp transitions necessary for transmitting digital information. For frequency modulation the responses are changes in frequency. The digital input signal shifts the phase, not the frequency signal, causing discontinuities, or non-smooth kinks, in the carrier's wave form. These discontinuities cause signal energy to be splattered outside the signal's channel. Part of the art of digital signal technology is the shaping of time transitions to keep the bandwidth response contained in the channel. To observe these effects it is necessary to view the time wave forms in their corresponding frequency spans.

Viewing the signal in the frequency domain

The frequency domain display in Figure 2.9 should be familiar to you. It is a frequency span, or spectrum, display of a portion of an NTSC cable television system. Frequency domain simply means the display of signal power versus frequency.

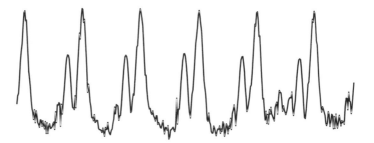

Figure 2.9. Several NTSC television channels in a frequency domain display.

Why use the frequency domain? It allows you to see signal power at each frequency in the spectrum and the general well-being of your system. You can also see signal responses that cause trouble to analog performance testing: interference, distortion, and noise. The primary test instruments for these measurements are the **spectrum analyzer, cable television analyzer,** and **signal level meter** (SLM).

What is in a signal?

How do you know what is in a signal?

As discussed, the sine wave seems to be a unique signal waveform; there are no other waveforms that go into the creation of a sine wave just the way the color yellow cannot be mixed from any other colors. The sine wave is the basic building block of all other signal waveforms. All other signals can be shown to be made up of a combination of sine waves of various frequencies, phases, and amplitudes. An example of this concept is the squarewave, which, at first does not look at all like a sine wave, except that it has a constant period. Figure 2.10 shows how a series of sine waves add to make up a square wave. Without getting into the mathematics behind this signal's composition, suffice it to say that the square wave is made up a series of sine-wave signals, the highest amplitude of which is a CW signal whose frequency is the same as the frequency of the square wave itself. The fundamental signal is the lowest-frequency signal in the waveform. The other sine waves are odd harmonics of the fundamental, that is, signals whose frequencies are 3, 5, 7, etc times the fundamental. These harmonics are smaller in amplitude than the fundamental. Only the odd harmonics are valid because they reinforce the periodic positive and negative square wave; the even harmonics are not present because their period goes against the symmetry of the square wave form.

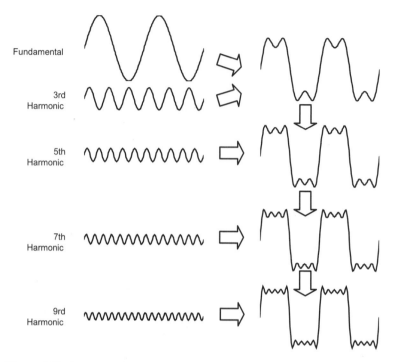

Figure 2.10. A square wave created out of odd-numbered harmonics.

Figure 2.10 shows the signal in the time domain as it might appear in an oscilloscope. The left column has the signal's sine wave components. The right column shows the construction of the square wave as more of the sine wave components are added to the composite signal. Adding more than a dozen odd harmonics converges the wave closer to a square wave as seen in Figure 2.11, where the approximated square wave is superimposed on top of the ideal square wave.

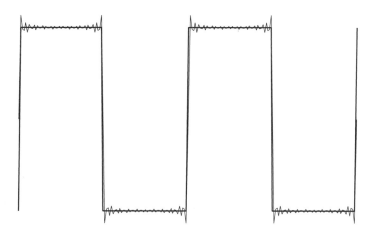

Figure 2.11. A square wave constructed of the first 18 odd harmonics of the fundamental.

Consider the frequency spectrum of this signal. The closer the waveform gets to a square wave, the more harmonics show up in the frequency domain. This means that when a signal changes direction abruptly in the time domain, as the square wave does at the corners, the signal's frequency spectrum is filled with harmonics. Said another way, the signal has its energy spread over a wide frequency spectrum because very fast action is necessary to construct the sharp changes in the square wave's amplitude. Fast action in an electric wave means high-frequency signals.

Signals must stay in their channel

An **oscilloscope** display and a spectrum analyzer display are combined in Figure 2.12 to demonstrate how frequency domain and time domain views relate to one another.

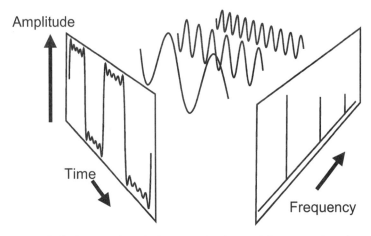

Figure 2.12. The relationship between the time and frequency domains.

Measurements in the time and frequency domains

The left face in the figure is the oscilloscope with the composite square wave at the front. The harmonic signals that make up the time domain signal are shown in the box, whose z-axis is frequency. The left face of the cube is the spectrum or frequency domain of these sine wave signals, where their amplitudes and frequencies are measured by the spectrum analyzer.

Signal frequency spectrum

The quicker as signal makes responses in the time domain, that is, kinks, corners, and wiggles in the time waveform, the higher the frequency content of the signal. Viewing the frequency domain of a signal with an instrument like a spectrum analyzer 1) shows the signal's highest frequency, and 2) helps predict the signal's required transmission bandwidth.

Knowing the frequency components that make up a signal allows you to determine the transmission bandwidth necessary to send this signal without distorting it. For instance, if the transmission bandwidth would only allow the fundamental and the first two odd harmonics, the receiver would see a signal that looked like the second waveform from the top on the right in Figure 2.10. It is not a clean square wave. Information about the signal shape has been lost by limiting the bandwidth of the transmission. But the smooth shape of this signal may be sufficient to send digital information.

To keep a signal in a specified channel bandwidth, it may be filtered. The combination of digital signal processing, digital modulation formats, and filtering are discussed in How an Analog Wave Becomes Digital Data, Chapter 3 beginning on page 33, and Viewing Digital Modulation, Chapter 6 beginning on page 77.

AM, FM and PM in the frequency domain

Figure 2.13 should be familiar. It shows a pilot carrier (a), an AM commercial station (b), a television FM sound carrier (c), a television PM color burst (e), and a digital signal.

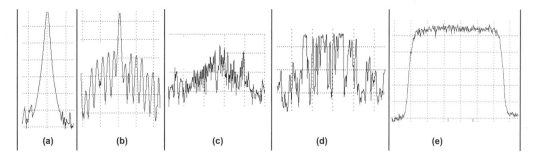

(a)   (b)   (c)   (d)   (e)

Figure 2.13. Frequency domain displays of common signals.

Analog measurements do not apply to digital signals

Table 2.1 relates the signal and modulation characteristics of broadcast and cable television signals to the types of testing done. It is important to note that in each of the characteristics across the table, the digital signal shares little or none of the characteristics or tests with any of the analog signals. Digital signal measurements are different from most analog television measurements done for analog performance.

Signal	Modulation	Signal characteristics	Frequency bandwidth tests	Amplitude tests	Modulation quality tests
Pilot tone during system installation	none	constant average, all power at one frequency	one frequency	absolute and relative power levels	residual FM, phase noise
Commercial AM station	amplitude	average power changes with modulation, signal carrier always present, bandwidth related to maximum modulation frequency	channel bandwidth	signal strength	bandwidth, percent AM, carrier power: prevent interference with other channels
Television video for luminescence	amplitude modulation in vestigial sideband NTSC or PAL	average power changes as modulation, limited lower sideband, carrier level changes with modulation	wide bandwidth required to reproduce vertical and horizontal sync pulses with sharp transitions	carrier level as peak of sync tips, adjacent channel levels, system sweep	signal power, in-channel frequency response, depth of modulation:
Television video for chromin-ance	amplitude and phase modulation placed in video sideband	chrominance amplitude is color saturation (intensity); phase is color hue (actual color)	phase information unchanged through system	color saturation	chrominance-to-luminance delay inequality (CLDI), differential phase, differential gain: subscriber color quality
Television aural carrier	frequency modulation	carrier at constant power	carrier frequency deviation	frequency deviation: power level relative to adjacent video carriers	frequency deviation, channel audio volume
Digital channel	digital	noise-like across channel	occupied bandwidth, adjacent channel power	total average power	modulation error ratio, error vector magnitude, bit, symbol and frame error rates

Table 2.1. Cable and broadcast signal characteristics.

Summary

Information is transmitted when the recipient of the signal cannot predict the content of the signal. A carrier signal is a CW signal that has no information of its own, but provides a stable, reliable signal to which the information is added. The information is contained in a signal of limited bandwidth, called a baseband signal. The process of modulation adds the baseband information to the carrier by changing the amplitude, phase, and/or frequency of the carrier. The common modulations are frequency modulation (FM), amplitude modulation (AM), and phase modulation (PM).

The best way to view modulation is to observe changes to the carrier in magnitude and/or phase relative to an unmodulated continuous wave carrier. A vector diagram shows magnitude and phase in a single circular, or polar, view.

Amplitude modulation puts the information on the envelope of the carrier. Analog video signals use a special type of amplitude modulation to transmit luminance and the synchronization signals required by the television receiver. Frequency modulation makes the frequency of the carrier change as a function of the baseband signal, leaving the amplitude of the carrier constant. FM is used for the sound of an analog television signal and is most commonly used in commercial FM stations. Phase modulation changes the phase of the carrier as a function of the baseband signal. PM is used to carry the chrominance, or color, information in analog television video.

Sharp changes in the baseband signal, such as required in the modulation of digital baseband signals, requires wider transmission bandwidth than the smoother signals in analog baseband. Generally AM and FM are more adept at handling these transitions than PM.

The frequency band characteristics of a modulated carrier are viewed in the frequency domain. The sharper a signal's baseband waveform, the more frequency bandwidth required to transmit its information. A square wave is made up of a theoretically infinite number of odd-numbered harmonics. The reduction of any of these harmonics at the receiver means that the signal cannot be reconstructed completely. A spectrum analyzer provides a convenient view and measures the quality of modulated signals in the frequency domain.

Questions for review

1. What is the one vital test for whether a signal contains information or not?

2. A carrier is
   a) a continuous wave signal with infinitely narrow bandwidth

  b) a continuous wave signal with very high frequency stability
  c) a continuous wave signal with very stable amplitude
  d) all of the above

3. A form of amplitude modulation is
  a) used for transmitting analog video chrominance
  b) used for transmitting analog video luminance
  c) changes the carrier amplitude as a function of the baseband information
  d) changes the carrier frequency as a function of the baseband information
  e) a and d
  f) b and c
  g) a and c
  h) none of the above

4. A periodic signal waveform can be reconstructed by adding together some number of CW signals. True or False?

5. A baseband waveform with sharp transitions is modulated onto a carrier without filtering or other signal processing. As the waveform transitions become sharper and/or more frequent,
  a) the modulated signal spreads over a wider bandwidth
  b) there is no difference in the signal's frequency response
  c) frequency spreading is dependent upon the type of modulation used
  d) a and c
  e) none of the above

Selected Bibliography

1. Paul J. Nahin, *The Science of Radio*, ISBN 1-56396-347-7, AIP Press, Woodbury, New York, 1996.
2. Harry Newton, *Newton's Telecom Dictionary*, 12th Edition, IBSN 1-57820-008-3, Flatiron Publishing, Inc., New York, February 1997.
3. Blake Peterson, *Spectrum Analysis Basics*, Hewlett-Packard Company, Application Note AN 150, Literature No. 5952-0292, Santa Rosa CA, 1989
4. Jeffrey L. Thomas, *Cable Television Proof-of-Performance; A Practical Guide to Cable TV Compliance Measurements Using a Spectrum Analyzer*, ISBN 3-13-306382-8, Hewlett-Packard Press, Prentice-Hall, Inc., 1995.
5. Robert A. Witte, *Spectrum & Network Measurements*, ISBN 0-13-030800-5, Prentice-Hall PTR, Inc., Simon & Schuster Company, Upper Saddle River, New Jersey, 1993.

3

How an Analog Wave Becomes Digital Data

An analog signal in time, that is, a waveform, is converted into digital data by interpreting the continuously varying analog signal into a series of numbers. This chapter teaches the fundamentals of this analog to digital conversion as the first step in the preparation of information for distribution as a digital signal in a broadband analog cable system.

What you will learn

- Converting an analog waveform to digital data
- Why is digital information easier for a machine to use?
- What is the difference between analog and digital transport?
- What is the relationship of bandwidth to digital signal transmission speed?

Digital data

Digital is discrete

Digital data are a set of values. A digital signal consists of digital data delivered over time, just like a waveform except that the values are from a specific predetermined set of discrete numbers. **Discrete** means that one value can easily be identified and distinguished from any other value. As an example of discrete numbers, take an empty egg carton and fill some of the pockets with eggs. Each cup represents a distinct value; the value with an egg is one, the value of an empty pocket is zero.

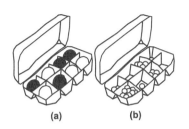

(a)   (b)

There is little ambiguity between values because the eggs are easy to see and count as in (a) in the drawing to the left. The value of the number represented by each cup is discrete, a one or a zero. A larger or more complex number is represented by the carton where the pockets are filled with a different number of marbles, as in (b), say from 0 to 9 marbles. Each pocket represents a single digit in a decimal number.

As long as the sender and the recipient of the carton agree on the order in which the carton's pockets are read, an 8-pocket carton can reliably represent numbers such as 14108496. Using only one item per pocket, such as eggs, the number would look something like 10010110. Clearly, there are more possibilities using marbles than eggs, so why choose eggs? Because the fewer values to choose from in each pocket, the easier it is to build and read.

The eggs represent values that are discrete, that is, easily separated or distinguished from each other. If you were asked to write down all the numbers from the carton with the marbles, you might have to unload some of the full cups to count the marbles; they pile up, making it hard to see all of them by sight alone. Even then some error, or ambiguity, would make your counting less accurate. But if the carton uses eggs instead, the number can be read almost at a glance. Reducing the range of values a number can have reduces the ambiguity and errors in communications for the same reasons.

Computers use counting schemes that allow only two values, stringing together unique sequences of these two values to represent specific numbers, just as in the carton with 8 pockets of eggs. Because this system is so important, it is given the name **binary**, or, consisting of two. The individual values, one and zero, are called binary **bits**.

Bits, binary, and bit stream	Bits are single digits in a binary number. A bit has a value one or zero. Binary means numbering with bits. A bit stream is a series of bits transmitted from sender to receiver for the purpose of communicating data.

Digital information is transmitted as a series of discrete and unique digital values. This flow of information is called a **bit stream**. The most graphic illustration of a bit stream in common digital technology today is the audio CD example, shown in Figure 1.1 on page 4. The pits and blanks represent a bit stream. These discrete values are only ambiguous if there is interference with the bit stream in the form of dirt or a scratch on the CD's surface. The detector tries to interpret a scratch as a value in the bit stream, adding errors and confusing the CD's receiver.

Binary: Two-value counting

Binary means something made up of two parts. In our daily use of numbers we use a system with ten value numbers, that is, the values 0 to 9. This is called decimal numbering. When a number is represented in a binary counting form, there are only two possible values. These are zero, 0, and one, 1. Computers use counting schemes that allow only two values, stringing together unique sequences of these two values to represent specific numbers, just as in the carton with 8 pockets of eggs. Because this system is so important, it is given the name, binary, or, consisting of two. The individual values, one and zero, are called binary bits. Any two characters can be used for the two values, as long as you recognize that one symbol is defined as the value 0, and the other is defined as the value 1. A byte is a binary number made up of 8 bits, such as 10011011.

Don't worry about learning the details of binary mathematics. It is rarely used in the practical day-to-day measurements of digital signals. Leave these computations to your test equipment's microprocessor.

Sending digital data

Computers store, calculate and transport digital information with ease and efficiency. Why? Because when they handle binary information

- simple two-state electronic switches can be used for storage and calculations
- the values 1 or 0 are easily stored, transported, and used in computation with little chance of misinterpretation

Digital travels well because of its discrete nature

A **digital signal** is digital data being transported over the entire electromagnetic spectrum; baseband, RF, microwave, and lightwaves. One reason a digital signal can survive the rigors of analog distribution media, such as cable, better than an analog signal is because of the digital signal's discrete nature. To illustrate, consider what happens when you shout down a long cement tunnel to a friend. In (a) of Figure 3.1 on page 36, an analog signal, in the form of shouted words, is distorted as the sound waves bounce off the pipe walls. The listener may have trouble understanding. But if the talker and listener agree to use Morse code, as in (b), using the talker's voice to sing a series of long- and short-duration tones, communication is more reliable. The listener has only to identify two types of signal values. The echo of the pipe interferes less with the binary vocalization than words made up of continuously varying tones and inflections.

Digital is robust

Complex forms of communication, such as voice transmission, are easily degraded because of the intricate algorithms used to interpret the signals. Listening to a voice communication is an example of a complex interpretive algorithm that translates the sounds into meaning. Digital code is simple by comparison, and is more tolerant of degradation before losing its meaning to the digital receiver-decoder.

Some digital communication systems use regenerative repeaters that read the digital signal, make each bit more discrete by making the variations between 1 and 0 more pronounced, and retransmitting the signal. This is like having a number of people in the pipeline of Figure 3.1 (c), each passing the digital message along like a bucket brigade.

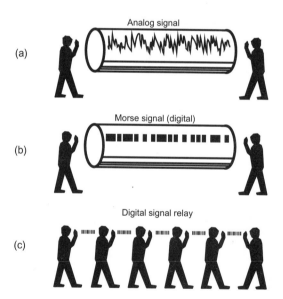

Figure 3.1. Analog and digital signals transmitted in analog transmission media.

Analog to digital signal processing

Signals are converted from one form to another form to enable their use, transport, or storage. Form change is called changing its **format**. Music is turned into digital information so it can be stored on a CD. **Signal processing** is a term used to describe signal conversion from one format to another. **Format** is the definition of the signal's specific, and often standardized, electrical characteristics.

Again, refer back to Figure 1.5 on page 11. The source signal is changed from an analog to digital format to make it suitable for modulation, which in turn, allows the signal to be transported as a channel in the system. The processes are reversed on the receiver side to return the signal back to a format for the subscriber's television or computer.

Analog to digital conversion, or **ADC**, is the first step in preparing a signal for distribution. To understand the ADC process, follow how the analog wave form, a squiggly line representing amplitude variation over time in Figure 3.2 (a), is converted to a series of numbers.

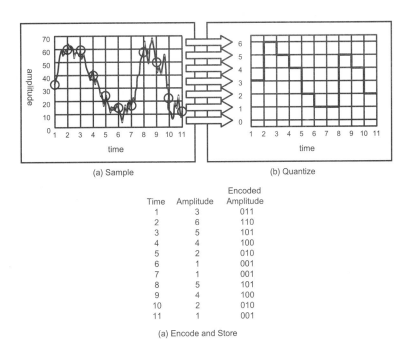

Time	Amplitude	Encoded Amplitude
1	3	011
2	6	110
3	5	101
4	4	100
5	2	010
6	1	001
7	1	001
8	5	101
9	4	100
10	2	010
11	1	001

(a) Encode and Store

Figure 3.2. Changing an analog waveform to digital information.

Snapshots for the analog wave form in (a) are taken at the time points 1 through 11, designated by the circles. This step is called **sampling** because only a few signal values, 11 samples in this example, represent the whole waveform. Each sample value is recorded as a number in a range permitted by the **quantizing** function. Quantizing is shown as an arrow between the (a) and (b) graphs. Each sample is quantized by assigning it a number corresponding to vertical level in which it appears. For example, the amplitude value being sampled at time 6 in (a), a value between 10 and 20 dB, is represented by the quantized value 1 in (b). The value is coded into a binary number, as shown in the third column of (c), and stored in memory for later use. These latter steps are called encoding and storage.

Each of these ADC steps impacts the signal's use, transport, and storage:

ADC = sampling, quantizing, encoding, and storage

1) **Sampling**: Taking snapshots of the analog signal fast enough to preserve its frequency response.
2) **Quantizing**: Turning the samples into amplitude values with enough resolution to preserve the waveform's changes.
3) **Encoding**: Turning the quantized samples into binary data that can be conveniently and efficiently stored and/or transmitted.
4) **Storaging**: Putting the digital data into memory for immediate or later use.

Analog to digital conversion

Analog to digital conversion, ADC, is signal processing that formats information for machines, like computers, modems, and fiber networks carrying digital signals. ADC is necessary to format information for digital modulation.

Each of these steps are explored in the following sections.

Signal sampling and quantizing

Sampling and quantizing determine how well the digital data represent the analog signal. Roughly speaking, sampling is responsible for restoring the original signal's frequency response, and quantizing determines the signal's amplitude integrity. Sampling and quantizing always result in permanently losing information contained in the original analog signal.

Figure 3.3 shows the effects of changing the sampling rate on the same analog waveform. The rate is represented by the value 1/T, where T is the sample period in seconds. Faster sampling means that T gets smaller, so the rate, 1/T, gets larger.

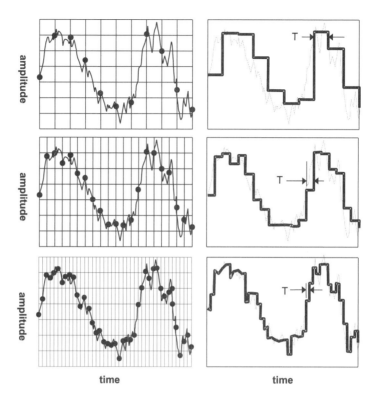

Figure 3.3. The effects of changing sampling rates.

The sampling rate is increased for each example down the page. The left column shows the sample points on the analog waveform as dots. The right-hand column shows the sampled waveform with the original analog wave as an overlay. As the sampling rate is increased, the sampled wave looks more and more like its analog source, tracing out many of the peaks and valleys missed by the lower sampling rates.

Nyquist criteria effects

To digitize a signal, a sampling rate of at least two times the highest frequency must be used. The transmission bandwidth of that digitized signal can be crudely calculated as two times the highest frequency of the analog signal. The Nyquist criteria are usually only used to determine a minimum sampling rate capable of preserving the signal's frequency content.

The relationship between the sampling rate and the frequency response of the waveform is given by the Nyquist criteria. The **Nyquist criteria** state that a sampling rate of two times or greater than the highest frequency of the signal is required to preserve the waveform frequency in the quantized data.

Sample rate, sample time	Digitizing higher-frequency analog signals requires higher sampling rates (smaller sample times). Sample rate in Hz = 1/(sample time in seconds).

For example, if the highest frequency in a baseband waveform is 4 MHz, then the sampling rate must be 8 MHz or greater to preserve the 4 MHz signal components. A sampling rate of 8 MHz means that a sample must be taken every 1/(8 MHz), or 125 ns. On the other hand, if the sampling rate is too low, not only do the waveform's high-frequency components get lost, but the fundamental waveform may be lost altogether, as in Figure 3.4. Here, the sampling rate on the left is adequate to trace the shape of the signal, but on the right the sampling is so slow that a new wave is generated by the sampled data. This effect is called **aliasing**.

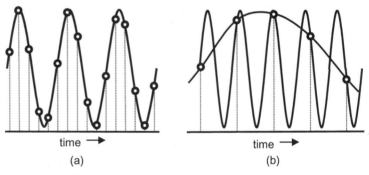

Figure 3.4. Sampling rate effect on two sine waves.

The higher the frequency content of a waveform, the more amplitude changes occur in a given time period. To capture these changes as data, the amplitude response must have finer-grain changes, that is, the quantization requires more values over the same amplitude excursion for faster changing amplitudes.

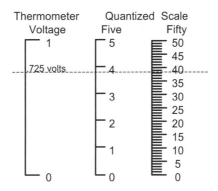

Figure 3.5. Two scales representing one volt.

For example, if a temperature gauge has a 0- to 1-volt output representing the thermometer calibrated range of 0 to 100° C, the precision to which you can read the temperature depends upon how many units are used to resolve the 0- to 1-volt scale. The voltage is read by a data acquisition device that displays the temperatures based upon the data's quantization. See Figure 3.5. The 1-volt range on the left is divided into 5- and 50-unit scales, representing the different data acquisition quantization levels. Temperature voltages are quantized, stored, changed into temperature units, and displayed.

If the thermometer output voltage is 0.725 volts, as measured by a precision voltmeter, the temperature would display as 0.725 x 100, or 72.5° C, the level shown by the dashed line. But the quantization of this value reduces the accuracy of the reading. With the quantization scale of 5 the readout could be either 3 or 4, depending upon how the rules used for data acquisition. For this example, assume the value chosen is the one closest to the voltage scale, 4. The output would show up on the readout as 100 x 4/5, or 80° C. For the 50-unit scale, the readout would add another significant decimal, for a reading of 38, for a readout of 100 x 38/50, or 76° C. The 50-unit scale provides a better representation of the actual analog value. Higher-unit quantization scales result in even better resolution of the temperature. The difference between the actual level and the quantized value is called quantization error. In the 50-unit quantization, the error is ±(76 - 72.5) = ±3.5° C.

Quantization error	The difference between a digital value and the analog value it represents due to the resolution limitations in the analog to digital conversion process. The smaller the quantization error, the better the digital signal reproduces the analog signal it came from.

Sampling and quantization work together to provide data that reproduce the original waveform. The effects of each are shown in Figure 3.6. An analog signal is sampled in the upper-left graph. Its data are a crude representation of the waveform. Improving the quantization is shown to the right, and improved sampling is shown below on the left. Combining both faster sampling and higher quantization provides the best waveform reproduction in the lower-right graph.

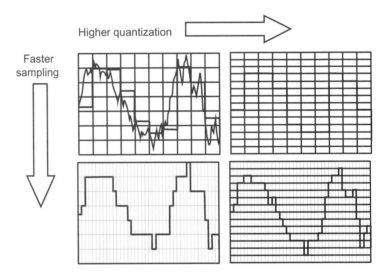

Figure 3.6. Effects of both faster sampling and higher quantization.

The waveform data are stored as binary data because computations and storage of digital information are efficient, fast, and usually cost-effective. The number of values required to quantize a waveform determines the number of binary digits required, just as the size of a decimal number determines the number of decimal places. For example, the decimal number 3410, requires 4 decimal places. Table 3.1

shows how to calculate how many binary places are required for different quantization values.

Number of quantization bits = n	Number of different quantization values, including zero = 2^n
0	not used
1	2
2	4
3	8
4	16
5	32
6	64
7	128
8	256

Table 3.1. Determining the number of binary bits necessary for quantization values.

The left column is the number of binary digits, and the right column shows the number of values that can be used for the number of digits. For example, if 16 values are required, including the number zero, the binary number requires 4 digits. The largest binary value for 4 binary digits, or bits, is 1111, a value of 15. Add the number zero for a total of 16 possible values in a 4-bit binary number. For some exercises in using basic binary math computations, see the Interval reference in the "Selected bibliography" on page 51.

How many bits?	The number of bits required for a specific digital application determines the quantization error and accuracy. The trade-off is the cost of storage and speed of computations.

The more bits computed and stored for an analog waveform's sample, the better the amplitude resolution and its accuracy. However, there is a point of diminishing return, where the quantization error is acceptable for the application, and/or the cost of computation and storage becomes prohibitive. One other factor impacts the quantization and sampling rate: transmission bandwidth.

If you work with a cable television system or a wide-area network, or WAN, you are already familiar with the concept **transmission bandwidth**. Channel bandwidth as well as the system forward and return bandwidths are illustrated in Figure 3.7. These are all called transmission bandwidths because they specify the spectral allocations for a signal or a set of signals. The transport medium is the cable and/or fiber optic cable.

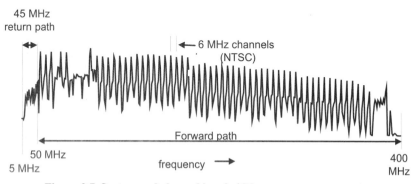

Figure 3.7. System and channel bandwidths.

Bandwidth

Bandwidth is the frequency range necessary to send a signal without degrading its quality. Bandwidth is also called transmission bandwidth, channel bandwidth, BW, and signal bandwidth. Pay attention to how the word is used, however, because it has many uses in communications. For example, bandwidth can refer to the total frequency range of your cable system, or the setting on a receiver's tuning frequency width. The term broadband can refer to the RF nature of your cable system but is also used in computer networks to describe the physical transmission medium for network communications.

Transmission bandwidth

Why not reproduce the studio-quality video directly into a digital signal for transmission to your subscribers? Aside from the high cost of the hardware and software to make these improvements, you would be limited by a much more important natural resource, frequency bandwidth.

Faster bit rate needs more bandwidth

The bandwidth required to send a signal is related to the amount of data in the signal and how fast information flows. The amount of data is related to sampling rates and quantization levels. The speed of the transmission, that is its data rate, is dictated by how fast the information has to get to the user. Increasing the number of bits sent over the same time period increases the required transmission bandwidth. This is a critical design criterion for system and signal design. Figure 3.8 is a cartoon demonstrating that for every increase in data rate or signal resolution there is a corresponding increase in required bandwidth. The vertical axis represents an increasing performance, such as sample rate increase and quantization error decrease. The transmission bandwidth for the changing performance is shown on the right.

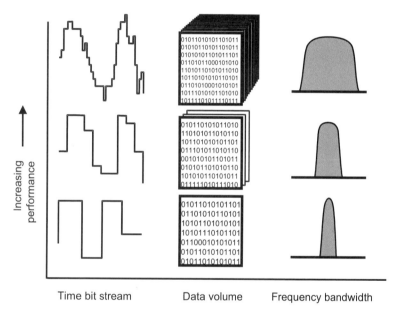

Time bit stream   Data volume   Frequency bandwidth

Figure 3.8. Effects of increasing digital performance on transmission bandwidth.

Transmission bandwidth is a limited natural resource

Bandwidth, whether in a cable, HFC, or over the air, is a limited resource. Transmission bandwidth, along with the time over which program and data is delivered, and plant length is one of the most expensive commodities in the design, installation, and maintenance of a communications system. In other industries, there are alternatives for handling transmission bandwidth, as the following box illustrates.

Hand-delivered bandwidth

From Andrew S. Tanenbaum's book *Computer Networks* comes this creative cost comparison of the bandwidth, time, and distance efficiency of media:

"One of the most common ways to transport data from one computer to another is to write them onto magnetic tape or floppy disks, physically transport the tape or disks to the destination machine, and read them back in again. While this method is not as sophisticated as using a geosynchronous communication satellite, it is often much more cost effective, especially for applications in which high bandwidth or cost per bit transported is the key factor. A calculation makes this point clear. An industry standard 8-mm video tape can hold 7 gigabytes. A box 50 x 50 x 50 cm can hold about 1000 of these tapes, for a total capacity of 7000 gigabytes. A box of tapes can be delivered anywhere in the United States in 24 hours by a package delivery service. The effective bandwidth of this transmission is 56 giga-bits/86400 sec or 648 Mbps, which is slightly better than the high-speed version of ATM (622 Mbps). If the destination is only an hour away by road, the bandwidth is increased to over 15 Gbps."

"For a bank with gigabytes of data to be backed up daily on a second machine (so the bank can continue to function even in the face of a major flood or earth-quake) it is likely that no other transmission technology can even begin to approach magnetic tape for performance. If we now look at cost, we get a similar picture. The cost of 1000 video tapes is perhaps 5000 dollars when bought in bulk. A video tape can be reused at least ten times, so the tape cost is maybe 500 dollars. Add to this another 200 dollars for shipping, and we have a cost of roughly 700 dollars to ship gigabytes. This amounts to 10 cents per gigabyte. No network carrier on earth can compete with that. The moral of the story is: Never underestimate the bandwidth of a station wagon full of tapes hurtling down the highway. "

A similar case could be made for comparing the competitors to cable television services, whether it be video tape rental, digital satellite broadcast, or Internet services.

Bits to bandwidth

How much transmission bandwidth is required to send a digitized waveform? At its simplest, the bandwidth has to be wide enough to contain the Nyquist sampling frequency. In reality, to make the most of the bandwidth for the type of data being transmitted, the bandwidth is set considering the signal conversion, encoding, and modulation processes. These latter considerations are covered in detail in the following chapters on signal layering and modulation.

Basic transmission bandwidth is tied to the Nyquist criteria. For example, consider a voice transmission whose maximum frequency is 4 kHz. Nyquist says that the sampling rate has to be two times this frequency, or 8 kHz. The transmission bandwidth is simply the sampling rate, 8 kHz. If the 4.25 MHz analog NTSC video signal were digitized using this same rule, the digitized signal would require at least 9 MHz of bandwidth. The comparative bandwidths for these analog and digital signals are shown in Figure 3.9. When the Nyquist criteria are not met, the signal being digitized gets badly distorted and may not be recoverable.

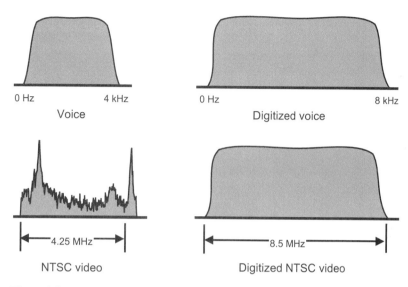

0 Hz   4 kHz
Voice

0 Hz   8 kHz
Digitized voice

|←——— 4.25 MHz ———→|
NTSC video

|←——————— 8.5 MHz ———————→|
Digitized NTSC video

Figure 3.9. Bandwidth requirements for rudimentary digitized voice and video signals.

Digitizing television channels for "digital" television is clearly not an efficient way to design a system. Each channel digitized in this way would take about two times its current bandwidth, reducing your cable system's television delivery by half!

Encoding, compression, and error correction

Much of the rest of this book shows how a digital signal's content and processing, even though it conforms to sampling rate and the Nyquist criteria, can contain so much more information than an analog signal in the same bandwidth. Clever signal processing technologies that encode formats, compress, error correct, and equalize make the transmission of quality digital signals possible. But first, let's define some terms you will need to understand these processes.

Bit rate

Error rate (or ratio) is a critical digital parameter

A digital signal's payload is the number of bits sent over time, called **bit rate**. The quality of the data is a comparison of the bad bits to the good bits sent over time, called **bit error rate** or **bit error ratio,** respectively. Bit rate is the number of bits sent per second, that is, in **bits per second**, or **bps**. Bit error rate (ratio), or **BER**, is a measure of the signal quality. It is covered in Digital Signal Quality, Chapter 8 beginning on page 127.

Bit rates use the same quantity adjectives that frequency units use, but instead of Hz, it is bits per second, abbreviated bps, as shown in Table 3.2.

Unit	Meaning	Example for bit rate
1	X1(times one)	10 bps = 10 bps
k, kilo	X1,000	15 kbps = 15,000 bps
M, mega	X1,000,000	100 Mbps = 100 million bps
G, giga	X1,000,000,000	1.5 Gbps = 1.5 giga bps = 1.5 billion bps

Table 3.2. Standard bit rate prefixes.

Communications transmission media, such as cable or optical fiber, has an associated transmission bandwidth and bit rate. The relation between bit rate and transmission bandwidth is dependent upon many factors, including

- type of signal information,
- coding format,
- acceptable bit error rate,
- and the quality of the transmission media.

Bit rates are limited by the transmission media, just as the frequency bandwidth of a cable is limited. Table 3.3 shows the upper limit bit rates for several media. Many transmission in these media are run at lower rates for reasons of reliability and quality.

Signal or Transport	Bandwidth	Maximum Bit Rate
Tin cans and string	<2 kHz more or less	Depends upon length and strength of string
Single plain old telephone system (POTS)	4 kHz	33.6 to 56 kbps
Single processed channel	6-8 MHz	30 to 40 Mbps
Composite digital video	9 MHz	135 Mbps
Twisted pair of copper wire	25 MHz	12 Mbps
Cable television system	450 to 750 MHz	450 to 3000 Mbps
Single laser optical fiber	20 GHz	10 Gbps
Bundled optical fiber cable	Multiply the above single laser	Multiply the above single laser

Table 3.3. Typical rates for digital traffic based on industry standards and experience.

Restoring the analog signal from the digital signal

For most digital signals, the conversion back to a useful analog form is done by dedicated receivers, like a television set-top box, a fiber to RF node module, or a data interface card in a networked computer. The transmitter and ADC usually encode enough intelligence, error correction, and raw data to make the transition back to an analog signal a comparatively mindless process at the receiver unless the signal has been corrupted by distortion or interference. Some deterioration can be fixed prior to, or at, analog to digital conversion, and others cannot. These digital signal characteristics and the impairments that cause the problems are the subjects of the next few chapters.

Summary

Analog signals are those that are continuously variable. Digital signals are composed of discrete quantities in a stream. All transmission media is analog, even though the signals sent may contain either analog or digital.

Analog waveforms are converted into digital data with a signal process called analog to digital conversion, or ADC. This conversion requires sampling of the analog waveform, quantizing the sampled values, encoding the values into a form suitable for storage or transmission, and storing the data for later retrieval or transmission. To preserve the analog waveform's frequency response, the ADC sampling rate must be at least twice the highest frequency in the analog waveform. This principle is called the Nyquist criteria. Higher signal fidelity also requires dividing the amplitude of the analog signal into a greater number of values. This process is called quantization. The quantized values are encoded, usually into binary numbers, for transmission or storage. The more values used for quantization, the more bits in each data point.

Transmission bandwidth is the most valuable resource in a cable system because it determines how much information, digital or analog, can be distributed. Higher data rates, for better signal quality, require wider bandwidths. The next few chapters cover the processes used to manage a digital signal's bandwidth while retaining signal quality.

Questions for review

1. When a signal is put through an analog to digital conversion, then a digital to analog conversion, some of the analog signal's information is lost. (True or False?)

2. What are the important parameters to set when converting an analog signal to a digital signal? (check all that apply)
    a) sample rate
    b) detection method
    c) filtering before the ADC process
    d) how much digital storage memory is available
    e) how the data can be processed
    f) near real-time processing of data is required

3. For digital representation of a signal to be accurate, the sample rate should be
    a) less than two times the highest frequency of the analog signal
    a) at least as high as two times the highest frequency of the analog signal
    a) equal to two times the highest frequency of the analog signal
    a) more that two times the sample period

4. Transmission bandwidth is a simple calculation based on the number of bits per second that have to be sent. (True or False?)

5. What determines the costs of transmission? (Which is most true?)
    a) the larger the bandwidth the greater the cost
    b) costs are proportional to the bandwidth divided by the time which they are used

c) costs are proportional to the time, number of bits, and bandwidth used

d) bandwidth alone determines the cost of transmission, especially in the cable system you own

Selected bibliography

1. *Interval*, the SCTE Newsletter, Society of Cable Telecommunications Engineers, Inc. *DigiPoints*, Exton, PA, November/December 1996.

2. Clyde F. Coombs Jr., *Electronic Instrument Handbook*, 2nd Edition. ISBN 0-07-012616-X. McGraw-Hill, Inc., New York, 1995.

3. Brian Evans, *Understanding Digital TV; the Route to HDTV*, ISBN 0-7803-1082-9, IEEE Press, Piscataway, New Jersey, 1995.

4. Daniel Minoli, *Video Dialtone Technology, Digital Video over ADSL, HFC, FTTC, & ATM*, ISBN 0-07-04-2724-0, McGraw-Hill, Inc., New York, 1995.

5. Andrew S. Tanenbaum, *Computer Networks*, 3rd Edition, ISBN 0-13-349945-6, Prentice-Hall, Inc., New Jersey, 1996.

4

Distributing, Layering, and Multiplexing

The interaction between a signal's composition, environment, and transport medium can be confusing. This chapter unscrambles the terminology and technologies involved in the distribution and composition of digital signals in an analog world.

What you will learn

- What helps a digital signal survive the trials of distribution?
- Why is digital signal layering?
- Why is a digital signal's content hidden from traditional time and frequency domain views?
- What role does signal multiplexing play in cable television system measurements?

To prepare a signal for distribution in a broadband analog system as a digital signal, it needs to be

- digitized and encoded,
- modulated onto a carrier,
- combined with other signals (optional),
- upconverted in frequency,
- and distributed to the user.

Each of these processes requires integration with the current distribution systems. This chapter provides some insight into the nature of a digital signal, and an overview of the technologies that challenge the cable industry as digital services are added.

Future distribution challenges

There is no doubt that future cable systems will host digital signals for entertainment as well as data retrieval. These system are capable of growing to meet the needs of subscribers who, not only want to be entertained digitally, but require the use of computer networks, such as the Internet. Figure 4.1 is a speculative overview of the evolution of systems towards those needs. Not long ago, as shown in (a), the cable system served primarily as a one-way analog entertainment system. The diagram in (b) reflects what is happening in the 1990s as digital entertainment adds to the host of analog channels, and the beginnings of connectivity to data through a computer network link. The extension of this is shown in (c), as the distribution system provides full digital entertainment and computer network interaction, integrating broadband analog and computer network attributes.

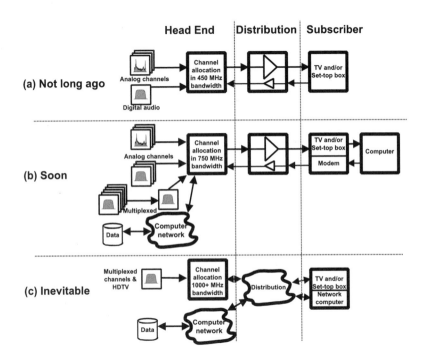

Figure 4.1. Cable television system evolution.

The future focus is on system reliability, and the ability to deliver and receive lots of data from subscribers and networks efficiently. For you to understand system and signal measurements, you need to recognize the attributes and jargon surrounding digital signal layering and multiplexing. This will help you in your installation and maintenance jobs. As additional communications services are added to your system, such as telephony, personal communications systems (PCS), computer network interfaces, and as yet invented digital technologies, system layering, signal layering, and multiplexing will play increasingly important roles.

Digital signals are different

Good design and maintenance of the design criteria are the reasons a system works well. Maintaining the standards established by the design depends upon your ability to monitor tell-tale parameters, such as errors, distortions, and signal quality. A digital signal cannot be measured with the same tools or procedures as an analog signal because it is "built" differently. One of these differences is **layering**.

What is layering?

The term comes from the computer networking software industry. Layering means designing software as a set of modules; each one performs a specific task and interacts only with adjacent modules. The restriction to two interfaces makes each layer simpler to design and maintain.

Each layer communicates to the layer above and below in functionality. Computer network layering is used to standardize data transmission while supporting different standards at each layer. Standards for data transmission in almost any medium, from twisted-pair to optical fiber, have layering to help keep sanity among designers, installers, test equipment manufacturers, and system maintenance staff. Design continuity is kept by carefully specifying the communications standards between each layer. The communications standards between layers are called protocol standards.

Layering	Layering is a software design architecture where each functional part, or layer, which performs a specific task needs to interact only with the layer above and below. "Above" usually means a task more toward the user application, and "below" means toward the input/output.

Layering also applies to the encoding used to package a digital signal. The uses and value for a digital signal are similar to those for a computer network. **Layering** is like a series of translations through which every piece of data that travels through a system must undergo. It is an organized way to separate a message from the media that sends the message. For a simple example, see Figure 4.2. The objective of this layered communication is to see that both people at the ends of the communications chain get their messages in English.

A message in English is sent by an American salesperson in Seattle to an American coffee broker in Bogota, Colombia. But the office staff in both locations speak Spanish. The top layer is the messaging at both ends of the line. The second layer translates the message into Spanish. The third layer creates a FAX in Spanish, and the fourth layer, the physical layer, is a FAX. The telephone line is the physical layer. At the receiver end, the message is processed in reverse order. At each layer there is expertise up one layer and down one layer. But the expertise of layers one or more removed does not have to be learned. For example, the salesperson at the coffee store does not have to learn Spanish, or even know how to use a FAX machine. The expertise at each level is critical, however. For instance, it is mandatory that the translators in Layer 2 know both English and Spanish equally well at the risk of having an order for 10,000 kilos of coffee beans be misread as an order for 10,000 pounds!

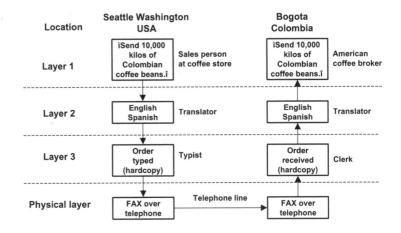

Figure 4.2. Layering of communications.

Digital signal layering is the encoding process that prepares the digitized waveform for transport and unwraps it at its destination. These processes include compression, error correction, and equalization. For details on these processes, see Error Correction, Equalization, and Compression, Chapter 7 beginning on page 99.

System and signal layering	A computer network uses layering to maintain a reliable and measurable data flow. A digital signal is layered to keep its information compact and protected from transmission problems. When a digital signal is sent over a computer network, the protected digital signal, already layered within itself, is sent through the computer network's layers just like any other data stream.

Digital signals wrapped within these layers can be sent by a number of different means, such as cable, HFC, or even a computer network. If the digital signal is sent by computer network, it is processed further by the computer network's layers. It is easy to get the concepts and the buzz words for these two distinct processes confused. In this book, as with your cable television system measurements, the focus is on digital signal layering.

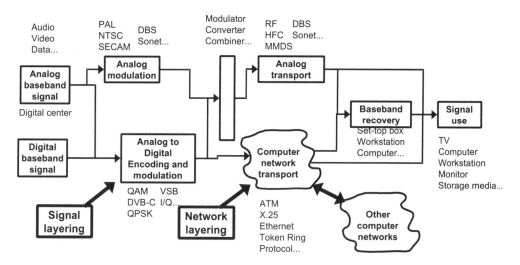

Figure 4.3. Distribution and processing of signals in a cable system.

Figure 4.3 shows the distribution and processing of a signal. The signal sources on the left are processed appropriately by the analog or digital blocks to provide a signal suitable for upconversion and combining onto the RF or computer network transport. Digital signals can use a system's proprietary computer network for distribution and interface with other computer networks. Signal layering is used in the encoding of the signal, and network layering is used for transport within and between the network "clouds." Each block has a sample of the terms, standards, and acronyms associated with its use and function to help you understand the relationships in a signal's life. For example, if you see the word protocol in an article, you know that the subject concerns the computer network, not signal layering.

Digital signal layering

Digital signal layering packages a signal with digital processes that compress, error correct, equalize, and modulate. Figure 4.4 shows the encoding that goes on at each layer. On the transmission side, the signal information is prepared for transportation. Upon arrival at the receiver, the signal is decoded in the reverse order, to retrieve the baseband information. The following chapters describe each of these processes, and their relationship to digital signal measurements.

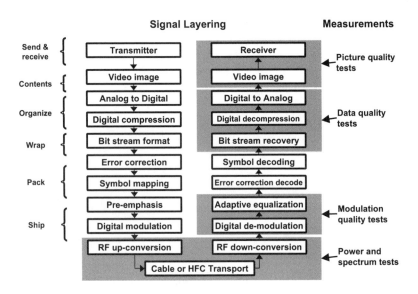

Figure 4.4. Digital signal layering.

The purpose of these layers is to protect the signal from transportation impairments. But the annoying effect of this protection is to hide the contents of the signal from viewers like you in your efforts to maintain the cable system. You cannot observe the quality of the signal and its contents along the signal path of Figure 4.3 without special test techniques and equipment. DVB-C, a set of standards for digital broadcast over cable from a consortium of European broadcasters and manufacturing companies, defines every layer in the process to keep suppliers of hardware, software, and system components compatible, and subject to the same compliance test limits and procedures. Standards are being aggressively created in the Americas for the same purpose.

Layering packages signals

The digital-analog allegory

Figure 4.5 is a cartoon to help you understand the differences in analog and digital signals being transported over a cable system. A signal converted to a series of ones and zeros, called a data stream, can be thought of as a series of white and brown eggs set into cartons specially built for shipping. The eggs are put in a specific order in each container to represent information that the recipient can read.

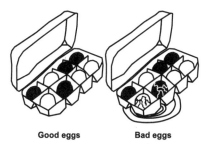

Good eggs   **Bad eggs**

If the eggs are broken on receipt, all information is lost. The recipient would only know that the information is bad, not how to repair it. Broken eggs are worthless.

The analog signal, a PAL or NTSC video signal, passing through the same system, is like a pot of spaghetti. Its contents are visible and available throughout its journey through the system. If the spaghetti gets cold, it can still be recognized and eaten.

Since the digital signal is potentially more fragile, it must be protected with a protective cover of packaging. The wrapping of the eggs in bubble pack and then in a box represents the layers of compression, error correction, and preemphasis that protect the signal's data from corruption from transportation. Just like a package in brown paper, the contents of the digital signal cannot be seen in the time or frequency domains.

For both the analog and digital signal, the final wrapping is the modulation and upconversion that puts the signal into its channel slot for transport to the subscriber. The RF/HFC truck takes the analog and digital signals in the same payload and unloads them at the same place, the set-top box. The digital signal needs unwrapping and unpacking; the analog signal can be consumed right away.

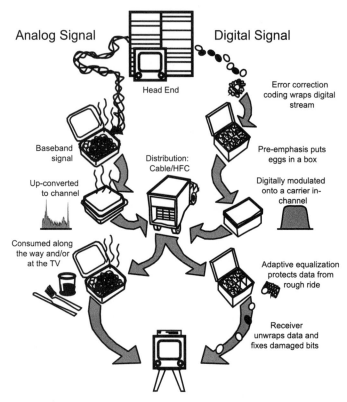

Figure 4.5. The analog and digital processing of video signals on their way to a television through a cable television system.

The following chapters explain the functionality of error correction, adaptive equalization, digital modulation, and the techniques for monitoring and measuring digital signal quality.

Combining digital signals

Multiplexing adds signals together

Digital technology facilitates the combining of several independent signals in one. Computer networks do this well. In fact, they are designed for that purpose. Broadcast and cable systems must also combine digital signals to improve services and reduce costs. **Multiplexing** is the ability of a communications circuit or channel to carry more than one independent signal, usually for the purpose of conserving frequency bandwidth. The earliest contributions in multiplexing technology came from increasing the transmission capability in the limited bandwidth of the telephone service's twisted copper wire pair. Multiplexing carries different meanings for different applications. To the cable television industry, it helps optimize the use of the cable's frequency spectrum for both analog and digital signals.

System or application	Multiplexing use	Jargon examples
Cable and HFC for cable television systems	multiple channels or bands transmitted over one cable or optical fiber	Frequency and wave division multiplex (FDM and WDM respectively)
Digital communications services	frequency division multiplexing (FDM) in the commercial digital telephone system	DS- levels, or digital services
Wireless cellular and personal communications systems transmissions	over-the-air transmission formats	Code and time division multiple access (CDMA and TDMA respectively)
Analog modulation	adding teletext, closed captioned, and other services to a TV channel's blank lines	teletext, closed captioned in NTSC and PAL
Digital modulation	putting more than one channel or data service in single digital television channel	QAM, VSB, QPSK, and COFDM

Table 4.1. Examples of multiplex use in communications systems.

Table 4.1 shows examples of multiplexing in a wide variety of communications industries. The sample jargon in the right-hand column can help you understand the use of multiplexing as your cable system begins to add communications services.

Multiplexing

Multiplexing is the process of putting more than one signal on the same communications circuit, usually to conserve frequency bandwidth. Multiplexing works with digital modulation techniques in time, frequency and code domains. "Division" and "multiple access" are words that usually indicate a multiplexing process is being used.

Multiplexing in the RF transport level is simply the addition of signals using a combiner. Techniques for adding independent data streams may not be quite so intuitive. As an example, consider yourself in a crowded room full of people talking to each other in groups. The fact that they are successfully talking is an example of multiplexing. What makes it possible for them to talk? Each person's voice and hearing frequency range is about the same. Individuals may be "tuning in" their group's voices because they are familiar, perhaps even speaking in a dialect or language that is closer to their own. The differences in pitch between group members also make it easier to listen. Or a person may use a pause in the general noise level of the room before conversing. Each of these techniques is used in digital signal multiplexing:

- Pitch differences for frequency division multiplex (FDM)
- Waiting for quiet for time division multiplex (TDM)
- Language or dialect differences for code division multiplex or code division multiple access (CDMA)

Multiplexing in time

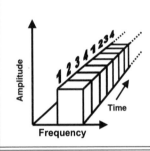

Time can be used to multiplex several signals together, while conserving both frequency and time domain. The wireless transmission format TDMA, or time division multiple access, divides each of its signals into equally long time slots. Each slot is repeated on a fixed or predictable periodic schedule. A single repeating slot sends the information of a single signal, so the TDMA channel can send as many signals over one frequency bandwidth as it has frequency slots. TDMA is used for the North American Digital Cellular, or NADC, cellular radio systems. In Europe, TDMA is used for the GSM and PCS cellular wireless systems.

Multiplexing in broadband cable and HFC systems

Most systems today use frequency or time multiplexing to combine more than one signal in a single transmission circuit. A multiple-channel cable system is a simple form of frequency multiplexing. Diplexers and combiners put all the channels on one "circuit" for distribution. There is no real conservation of the spectrum because all the channels occupy the same bandwidth all the time, but they stay out of each other's way. Figure 4.6 shows this graphically. The channels are operating over time, represented by the frequency spectrum profiles marching in time back into the page.

Spectral efficiency means keeping the most signals available for your customer at all times in the least amount of bandwidth. Figure 4.6, illustrates a simple example of frequency multiplexing you probably use in your system. The channels A, D, and E go off the air at 1:00 AM, so that other channels, not available during normal hours, can be put in their place. This is represented by the shaded portions of the channel rows after 1:00 AM, when the alternate channels are substituted.

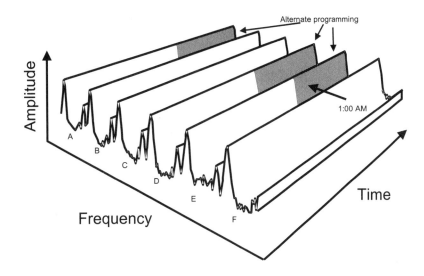

Figure 4.6. A cable system example of multiplexing in the frequency domain.

Multiple access, time division, and frequency division

These terms can be confusing, especially when they all relate to multiplexing. Multiple access means that multiple users of a system can access each other when they want. If they use common communication circuits then multiplexing is at work. Multiplexing is a way for multiple signals to operate over the same communication line. An example of multiple access is a group of computer users who can contact and send information to each other at any time.

Frequency division means that several signals occupy specific frequencies at the same time. Time division means that several signals are sent at the same frequency or over the same frequency band in alternating time periods. See "Multiplexing in time" on page 62.

Return path spectrum conservation requires

The return path is used for high-speed cable modem links with computer networks and other services through the system's head end. In most systems the return path is restricted to between 5% to 10% of the total system bandwidth. Multiplexing is required to allow as many subscribers on-line as possible.

Summary

Layering and multiplexing are integral processes to baseband signals which are digitized, modulated, converted, and distributed. Often the concepts and their uses are confused by the many terms, acronyms, and jargon in the trade literature.

Both analog and digital signals are distributed through RF and microwave systems. Digital information is transported on computer networks. Although computer networks are becoming part of cable systems, this book does not cover their design and use, except to help you distinguish between computer network and other digital communications techniques.

The techniques for transporting digitized information and signals include layering and multiplexing. Layering is a software architecture that insures an organized procedure is followed for data flow through a computer network. Layering principles are also applied to the packaging of digital signals for compact and robust transport. Modulation formats include layering definitions for digital signals.

Multiplexing is the process of putting more than one independent signal onto a single transmission line or frequency bandwidth. This process occurs at many communication levels, including cable systems, as the combining and diplexing of channels. In wireless communications multiplexing is implemented as frequency-, time-, and code-multiple access formats. In computer networks, multiplexing combines bit streams.

Questions for review

1. Name the four steps to distribute an analog baseband signal as a digital channel.

2. Layering is a technique in the design and construction of computer network and digital signal technologies. True or False?

3. Digital signal layering (select all that apply)
   a) protects a signal from transmission path problems
   b) is specified by some signal formatting standard
   c) is always part of a computer network layering standard
   d) none of the above

4. Match the analog/digital system building block, listed a-f, with the acronym or term g-l, that more closely belongs to its technology.
   a) Analog modulation
   b) Computer network transport
   c) Digital formatting
   d) Analog transport
   e) Analog modulation
   f) Baseband recovery
   g) protocol
   h) QAM
   i) X.25
   j) PAL
   k) set-top box
   l) up conversion

5. Multiplexing
   a) is a technique for putting more than one signal on a single
      electrical line
   b) is a general term for combining signals in broadband systems
      as well as in single channels
   c) helps conserve frequency bandwidth
   d) all of the above

Selected Bibliography

1. Harry Newton, *Newton's Telecom Dictionary,* 12th Edition, IBSN
   1-57820-008-3, Flatiron Publishing, Inc., New York, February
   1997.
2. Andrew S. Tanenbaum, *Computer Networks*, 3rd Edition, ISBN 0-
   13-349945-6, Prentice-Hall, Inc., New Jersey, 1996.
6. Matt Trezise, "Testing Digital Video Set Top Boxes and Networks," Internal Training Presentation, Hewlett-Packard,
   Queensferry Microwave Division, Marketing Department, February 1996
7. Ian Wright, *HP 8594Q QAM Analyzer Product Note*, Hewlett-
   Packard Company, Literature No. 5965-4991E, Santa Rosa CA,
   1996.

5

Introduction to Digital Modulation Formats

Digital baseband information is modulated onto a carrier for transmission, just as the modulation of analog information. Equally important in both cases is the need to select a modulation technique that keeps the baseband information safe from transport impairment while transporting the information efficiently. This chapter describes the general attributes of popular digital modulation formats, their uses, strengths, and weaknesses before digging into the details of how they work.

What you will learn
- How digital modulation formats differ
- Digital modulation format strengths and weaknesses
- What is the relation between multiplexing and modulation?

There are already many modulation formats, with three or more letter acronyms to represent their names. Each one has its own set of strengths and weaknesses. Modulation formats are being created and modified at a rapid rate to satisfy applications for business, personal use, and entertainment. The result is dedicated formats which deliver more information in the same bandwidth. Many of these format acronyms are plotted in Figure 5.1. The plot shows the increasing modulation efficiency over the years. Modulation efficiency, the ability of a format to carry more information in the same bandwidth, brings with it increased complexity, that is, requiring more expensive and technically challenging hardware.

Limits of modulation efficiency	As more information is put into the same frequency bandwidth, the complexity of the equipment increases and the communication channel requires more protection from noise and distortion.

The familiar analog modulation formats; AM, FM, and PM, could carry digital information, but their modulation design restricts their transmission bandwidth. They are not very efficient. Frequency shift keying, or FSK, was one of the first FM-like modulation formats created for transmitting digital baseband. The invention of the I/Q modulator/demodulator led to a whole new family of modulation formats, starting with QPSK and QAM, that increased efficiency and softened the line between modulation and multiplexing. I/Q modulation is the subject of Chapter 6 beginning on page 77.

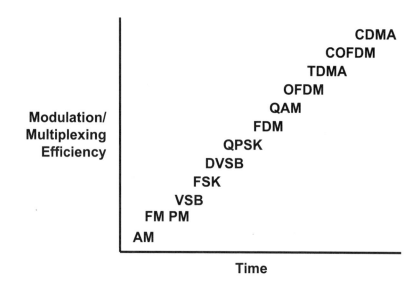

Figure 5.1. Digital modulation and multiplexing format efficiency over time.

Modulation and multiplex formats

Modulation and multiplexing work together

Digital modulation and signal multiplexing techniques work hand-in-hand to make the best use of both. Multiplexing uses modulation to send information. Usually a multiplexing technique lends itself to the use of a specific type of modulation format, and vice versa. Table 5.1 shows a number of multiplexing techniques, the modulation formats used with them, and examples of their use. Several of these multiplex types are currently used in personal communications, such as cellular radio, but derivative forms are beginning to show up in the communication of data over cable systems.

Future applications in the cable television and communications industry will add these, or variations of these multiplexing formats to enable applications yet-to-be invented. Today's cable television workhorse modulation formats, shown at work in Figure 5.2, are QAM, QPSK, COFDM, and VSB (or DVSB for digital VSB); system performance and measurement needs are looked at in detail.

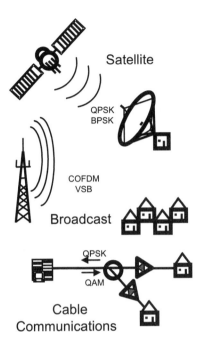

Figure 5.2. Popular uses of QAM, QPSK, COFDM, and VSB.

Modulation formats each offer different strengths

At this writing the world's digital modulation format standard for cabled network distribution is 64-QAM and 256-QAM. VSB and COFDM are the standards for broadcast because of their robust performance in environments of bouncing and delayed signal paths, called **multipath**, and the **fading** of signals due to atmospheric and weather conditions. In many markets, conversion between VSB and QAM may be necessary to enable cable systems to carry the broadcast programming, although in North America, this issue, called the must-carry policy, is far from settled. Governmental regulatory agencies may require cable systems to use the terrestrial modulation format as transmitted over the air. For example, the U.S. Federal

Communications Commission (FCC) may require cable systems to carry the original VSB signals under their "must-carry" rules.

Acronym	Description	Modulation	Use	Advantages	Disadvantages
FDMA	Frequency Division Multiple Access	AM and FM	Commercial radio	Simple	Inefficient use of frequency spectrum
TDMA	Time Division Multiple Access	Pi/4 DQPSK, GMSK	Wireless cellular and PCS (personal communication systems)	Many channels available	Complex management of modulation data access
FDMA/ TDMA	Frequency Division Multiple Access/Time Division Multiple Access	GMSK, pi/ 4DQPSK	GSM, PHS, PDC personal communications standards	Many channels available	Very complex management of modulation data access
CDMA	Code Division Multiple Access	QPSK, OQPSK	Wireless cellular and PCS (personal communication systems)	Immunity to interference of all kinds	Requires good signal-to-noise, complex multilayer decoder
OFDM	Orthogonal Frequency Division Multiplex	QAM (16, 32, and 64)	Cable television return path	Robust in cables with high impulsive ingress	Complex modulation integration
COFDM	Coded Orthogonal Frequency Division Multiplex	thousands of individual QAM carriers spread over the frequency band	Terrestrial video and audio transmission DVB-T	Very robust in high multipath environments, frequency spectrum conservation through a signal frequency network (SFN)	Very complex modulation and demodulation hardware

Table 5.1. Familiar multiplexing formats.

Many of the multiplex techniques and modulation formats in Table 5.1 are now used for the cellular and personal communications industry. As the cable industry begins to take more communications services, these formats and techniques will become integral to cable systems. Even though the discussion of many of these techniques goes beyond the scope of this book, your recognition of these terms and acronyms is valuable in understanding the wealth of technical articles, marketing promotions, and training material in the cable industry.

QPSK and multiplexing may combine in return path

The primary use of combined digital modulation and multiplexing is in return path communications, where interference and noise are at their worst. To make the most of the limited return path frequency bandwidth found in most cable systems today, single-carrier QPSK may be used in conjunction with code division multiple access multiplexing, CDMA. This combination process can spread each signal's data stream over the frequency band, making it less susceptible to the higher levels of interference and noise.

Digital modulation formats

Format efficiency, the ability to put more data in less bandwidth with acceptable reliability, directly impacts the cost of system operation. High financial returns await the system operators who have the most reliable product of bandwidth, time, and subscriber base. Modulation formats are only part of the story; error correction and equalization are mandatory for the reliable transmission of digital signals. These topics are covered in future chapters. Table 5.2 describes the popular modulation formats for cable television and broadcasting, with their advantages and disadvantages.

Acronym	Description	Use	Advantages	Disadvantages
AM, FM, PM	Amplitude Modulation, Frequency Modulation, and Phase Modulation	Commercial radio broadcast, private mobile, citizen's band, and cable systems	Inexpensive transmitters and receivers, small bandwidth requirements, proven suitability for the quality required	Not as spectrally efficient as more complex forms of modulation
PAL, NTSC	Phase Alternate Line (television), National Television System Committee	Commercial television broadcast and cable systems	Inexpensive transmitters and receivers	Quality lower than digital for most consumers, inefficient use of bandwidth
QAM	Quadrature Amplitude Modulation	High capacity microwave digital radio, cable modems, digital broadcast, digital cable transmission DVB-C	High spectral efficiency	Poor quality with low signal-to-noise ratio
VSB and DVSB	partially-suppressed-carrier Vestigial SideBand	North American broadcast digital television, modems	Robust carrier and symbol clock recovery in presence of multipath	High peak-to-average ratio, which makes high power transmitter design difficult
QPSK, BPSK	Quadrature Phase Shift Keying and Binary Phase Shift Keying	Low capacity terrestrial microwave digital radio, direct to home broadcast satellite, cable modem return path DVB-S	Robust in low signal-to-noise ratio	Poor use of frequency spectrum

Table 5.2. Analog and digital modulation formats for cable television systems.

There is no way to predict the mix of multiplex and modulation techniques. However, for video transmissions, compression is undoubtedly going to be an MPEG format or variation. See Error Correction, Equalization, and Compression, Chapter 7 beginning on page 99, for the impact of compression on your measurement needs.

QPSK and QAM

QPSK and QAM are closely related

QPSK and QAM are related by modulation and demodulation technologies, and, therefore, use the same measurement parameters and techniques. QPSK is the simpler of the two formats, transmitting 2 binary bits at a time. QAM is capable of transmitting 4 or more bits at a time. As with any digital modulation format, the higher the number of bits, the more susceptible the signal is to interference and noise, especially when the signal has not been encoded for error correction. Advances in coding and multiplexing allow the use of these formats for new uses. This will continue as long as there is a need for fast, reliable data streams.

The direct broadcast of television signals from satellites, DBS, uses the QPSK format because it is more resistant to noise and other interruptions over the long transmission distances. A tight, well-maintained cable system can provide the higher signal-to-noise ratio required of QAM. QAM, in its various forms, is roughly four to six times more efficient than QPSK.

Translating from one format to another

As information comes to the cable system from a variety of sources, it is often necessary to translate between modulation and multiplexing formats. One such device is the integrated receiver/transcoder, or IRT, which is used to receive digital signals via the satellite. The IRT converts downlinked satellite video signals from the QPSK-modulated carrier to a QAM carrier without changing their content.

Signals that are required to be carried in cable systems from over-the-air channels, the so-called must-carry signals, may require the cable system to translate between VSB and QAM at some future time. Although this seems like a roundabout method for satisfying the must-carry rules, adding VSB to the cable system may be an even bigger challenge because of the set-top box receiver requirements.

QPSK in cable systems is preferred for signal transmission in the noise-plagued return band between 5 MHz and 40 MHz. This format is more immune to the ham radio signals, citizens band transmissions, and various sources of impulse noise that are often found in the return path. Because of this, QPSK was chosen as the specification for the return path (from modem to head end) signal flow by the North American Cable Television Laboratories Data-Over-Cable Service Interface Specification (DOCSIS). The robustness does not come for free, however. QPSK can only achieve data transfer rates up to 10 Mbps.

There are a number of "flavors" of QAM, each with its own designation. Examples are 16-QAM, 64-QAM, and 256-QAM. The higher the number, the more information that can be sent in the same amount of time. For example, one high-definition television (HDTV) signal can be sent with 64-QAM in a 6 MHz bandwidth. A 256-QAM signal can carry two HDTV signals in the same bandwidth. The more complex a format, the more error-prone it becomes to transmission impairments; the higher-order QAM requires better signal-to-noise ratio to reduce errors. As mentioned, a lot of design and engineering has gone into QAM encoding to improve its use of available bandwidth.

VSB and DVSB

VSB used for over-the-air broadcast

VSB, which stands for **vestigial sideband**, has been adopted in North America for digital television broadcast over-the-air. There are concerns that digital VSB, sometimes referred to as DVSB, may be an island in a sea of QAM television applications. Europe has adopted COFDM for over-the-air transmission. At this writing VSB has not been added to the host of modulation formats in the set-top boxes which prepare the television signal for reception on analog television receivers. VSB may be added to television receivers in the U.S., however.

Summary

Modulation formats play a key role in the design and maintenance of a cable system. Different modulation formats offer different benefits for these applications. Format efficiency ties directly to the cost of system operation. The complexity of digital signals has risen to meet the demand of communications.

Digital modulation and signal multiplexing techniques work hand-in-hand to make the best use of both. Variations of FDM, TDMA, and CDMA multiplexing use various forms of these digital modulation formats.

The choice of modulation and multiplexing formats is far from settled. Currently in the U.S. QAM is the choice for transmitting digital signals in the forward cable path. QPSK, which is less efficient, but much more hearty in noisy environments, is the choice for the digital return path in cable. This could change as improvements to encoding and multiplexing are made. QPSK is also used for satellite transmissions because of its survival in broadband noise. VSB is the standard for digital over-the-air broadcast because of its robustness against fading and multipath. In Europe, COFDM is used for terrestrial digital broadcast and QAM is used for cable.

Questions for review

1. Name the three most-used digital modulation formats in North America, and their full names.

2. Multiplexing is often combined with digital modulation in communications systems. True or False?

3. Name the digital modulation format most suitable for each application:

    a) Over-the-air broadcast

    b) Return path digital communication

    c) Forward path cable broadcast

    d) Satellite direct broadcast

4. Which format is least susceptible to low signal-to-noise ratios?

    a) 8-QAM

    b) 16-QAM

    c) 32-QAM

    d) 64-QAM

    e) None, they are equally susceptible

Selected bibliography

1. "CED Cable Modem Deployment Update," *Communications Engineering & Design, Chilton Publications*, New York, March 1998.

2. *Interval*, the SCTE Newsletter, Society of Cable Telecommunications Engineers, Inc. *DigiPoints*, February 1998.

3. Michael Adams, "The Move to HDTV, Issues to Think About," *Communications Technology* magazine, Communications Technology Publications, Inc., April 1998.

4. Ken Freed, "QAM Delivery," *Broadcast Engineering*, Intertec Publishing, Overland Park KS, February 1997.

5. Ron Hranac, "DTV or Not DTV? That Is the Question," *Communications Technology* magazine, Communications Technology Publications, Inc., February 1998.

6. Louis Libin, "The 8-VSB modulation system," *Broadcast Engineering*, Intertec Publishing, Overland Park KS, December 1995.

7. Lawrence W. Lockwood, "VSB and QAM," *Communications Technology* magazine, Communications Technology Publications, Inc., December 1995.

8. Kenneth H. Metz, "Going Digital? Think Bit Error Rate," *Communications Technology* magazine, Communications Technology Publications, Inc., June 1997.

6

Viewing Digital Modulation

Digital modulation puts information onto a carrier, just like analog modulation, except the baseband is digital, not analog. The modulation techniques are significantly different from analog modulation techniques; they require different measurement parameters and test equipment. This chapter introduces digital modulation formats and their measurement parameters.

What you will learn

- How does information get coded into a digital signal?
- What is a symbol?
- What do in-phase and quadrature mean?
- What vector, constellation, and eye diagrams tell you
- The effect of filtering on the digital signal

A **digital signal** is a carrier which has had a digital baseband signal modulated onto it using **digital modulation**. Digital signals send information through the forward and return path of the cable system, just as analog signals do. Figure 6.1 shows the influences that the digital modulation format has in the overall measurement and parameter map of a cable television system, along with some of the test equipment used for monitoring signal quality. The modulation format has a direct bearing on the modulated signal's quality, dictating the quality test parameters and measurement instruments. The modulation quality plays a role in the quality of the data bit stream, and thus upon the digital channel's delivered quality.

Comparing analog and digital video signals

If you are familiar with cable television system test, the pictures in Figure 6.2 will be familiar. They are the frequency domain displays of NTSC and PAL television channels. In the frequency domain, you can easily pick out the video, audio, and color signals. And, because the signals are constantly moving and changing, you can tell something of the nature of the signals being transmitted: whether the picture is changing, if the audio's voice or music, and if the video signal is color or not. You can also see some signal quality characteristics, such as the video to audio signal strength ratio, and whether gross noise or distortion is present.

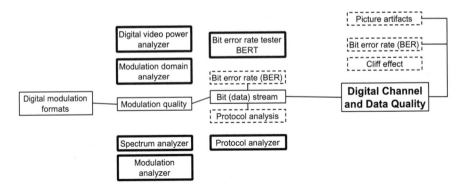

Figure 6.1. Signal flow involving modulation formats.

If cable television analog measurement procedures are not familiar to you, you can learn about them from the Thomas bibliography in reference at the end of this chapter.

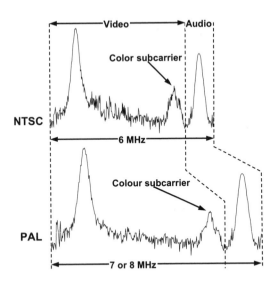

Figure 6.2. Analog television channels in the frequency domain.

You can't tell what's in a digital signal by looking at it

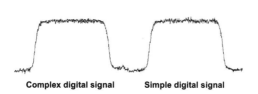

Complex digital signal   Simple digital signal

Generally, you can not tell what is inside a digital signal by looking at it. About all you may be able to tell from the frequency spectrum of a digital signal is the general type of signal, if it has gross distortion sidebands, if the passband has flatness distortion, and whether it is set to a generally suitable power level compared to neighboring analog signals. The reason for the veiled nature of the digital signal lies in the layered nature of its encoded content and its modulation scheme. To understand how these digital modulation formats are capable of sending information, it is necessary to study the tools that observed and analyze them. The first of these is the vector diagram.

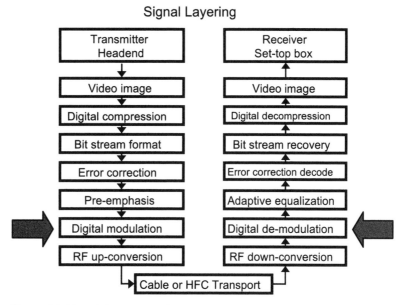

Figure 6.3. The digital modulation layer in digital signal processing.

Viewing modulation with vector diagrams

Chapter 2, Modulation and Frequency Management, demonstrated how to view analog modulation content. With more complex modulation, where phase, frequency, and amplitude may all be changing at once, such views are impractical and insufficient for the measurements required for digital modulation. For example, if amplitude, phase, and frequency are plotted on separate x-y plots, the variations of each are easy to see. This is shown in Figure 6.4, where the square-wave-like changes in each parameter over time represent digital values.

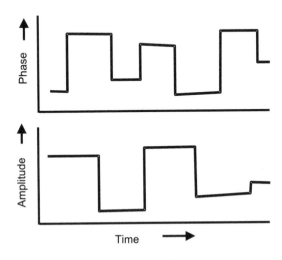

Figure 6.4. Digital modulation as a function of phase and amplitude.

Combining signals puts more information in the same bandwidth

If these separate signals can be modulated, or multiplexed, onto a single signal, more information can be packed into less bandwidth. To view changing phase and amplitude at the same time, a plot called a **vector diagram** is used.

A **vector diagram** is a polar plot of a voltage that has magnitude and phase. It is used to show the **changes** to the carrier as it spins, even though the changes occur simultaneously. A carrier can be represented by a vector diagram, but the result is a boring picture, since, by definition, a carrier does not vary its amplitude, phase, or frequency. To understand how a vector diagram presents modulation, recall the rotating stick in Figure 2.1 on page 16. The shadow of the tip of the rotating stick projects a time plot of its amplitude. To eliminate the rotation of the vector while observing the modulation, we can rotate right along with it. This is illustrated in Figure 6.5, an imagined contraption, where a TV camera spins only at the speed of the carrier vector. The measurement device can watch the changes to the carrier without being affected by the carrier itself. If the picture is transmitted to a TV, the TV screen only shows changes in phase and amplitude of the carrier vector.

Vector diagram	The vector diagram is a polar plot, that is, a plot of the amplitude and angle of a signal compared to the carrier. If there is no modulation, the diagram is static. Only changes of phase and/or amplitude are displayed.

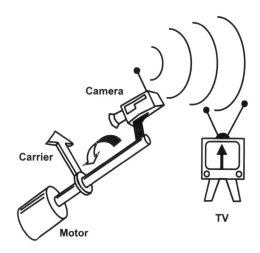

Figure 6.5. An electromechanical version of a vector diagram's source.

Scalar vs. vector amplitude vs. magnitude	Scalar and vector are mathematical descriptions of a value. Scalar means that parameter is defined by a single value, usually a length. Vector means that the parameter has both a length and a direction. In electrical parameters, especially voltage, direction is usually given as phase. The length of a wooden stick is a scalar value. The same stick is represented as a vector if both its length and the direction it is pointing are given. A polar vector's amplitude is called magnitude. Magnitude is an absolute value, meaning it has no + or - sign.

Figure 6.6 (a) shows the electronic version of the vector diagram. This diagram is also called a **polar diagram**, because, like a view of Earth from above one of its poles, the planet rotates about the center. The center of circular diagrams is usually called the **origin**. Amplitude modulation, shown in (b), appears as the vector moving in and

out at a fixed angle. The vector does not rotate because amplitude modulation does not change phase. For phase changes, (c), the movement of the signal relative to the carrier is an arc along the diagram's circumference. A combination of amplitude and phase change is illustrated in (d), where the signal's amplitude and phase change simultaneously. But such a diagram is not sufficient to show the changes in digital modulation, where data changes are represented by the movement to fixed points in the polar realm. The I/Q diagram and I/Q modulations schemes satisfy this need.

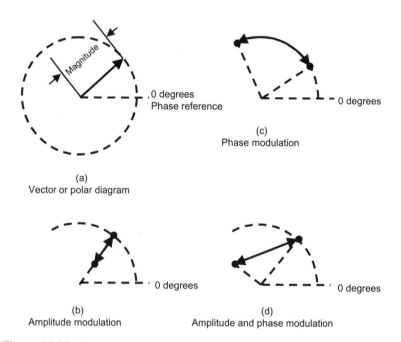

Figure 6.6. Viewing analog modulation with a polar diagram.

I/Q modulation

Modulating a carrier in amplitude and phase requires unique modulator and demodulator circuits. This modulation technique separates a vector signal into its so-called in-phase and quadrature elements, and treats each as if they were separate scalar parameters, just like the two time plots in Figure 6.4 on page 80. **In-phase**, also called **incidental**, means that the observed vector is coincident with the carrier, defined by traditional engineering notation, a minute hand from the origin pointing 3:00 o'clock. Quadrature means 90° away from the incidental direction, by convention counter-clockwise from the incidental vector. These are called **I/Q** for **incidental** and **quadrature phase**.

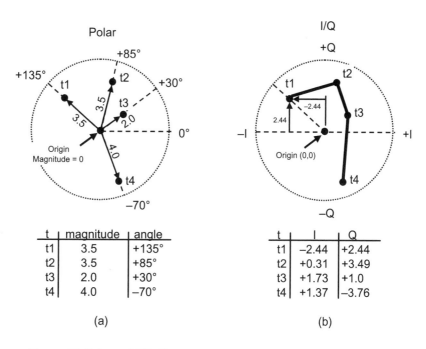

t	magnitude	angle
t1	3.5	+135°
t2	3.5	+85°
t3	2.0	+30°
t4	4.0	−70°

t	I	Q
t1	−2.44	+2.44
t2	+0.31	+3.49
t3	+1.73	+1.0
t4	+1.37	−3.76

(a)   (b)

Figure 6.7. Polar and I/Q diagrams.

Figure 6.7 is a graphical comparison between the parameters of a vector diagram and an I/Q diagram. In (a), a vector moves clockwise over time, t, as shown by the designations t1, t2, t3, and t4. Each position is recorded in the table below the drawing using the magnitude and angle of the vector. In (b), these same vector positions are given I and Q magnitudes just by observing the horizontal distance (±I) or the vertical distance (±Q) from the origin. Rather than a signal made up of angle and magnitude, it is defined in terms of an I and Q magnitude; the polar diagram points are described by Cartesian, or x-y, points. Instead of x-y, the terms used are I-Q.

I and Q values change independently of one another because they are 90° out of phase with each other. This relationship between I and Q signals is called **orthogonality**, which means at right angles. Two completely independent signals can be modulated onto this carrier, in a process similar to multiplexing. The implementation of this technique is in the modulator and demodulator box below.

I/Q modulator and demodulator

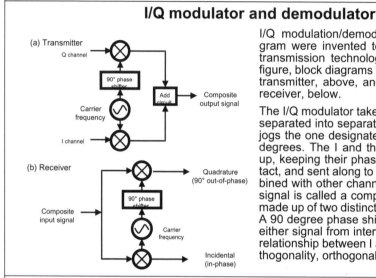

(a) Transmitter
Q channel
90° phase shifter
Add circuit
Composite output signal
Carrier frequency
I channel

(b) Receiver
Composite input signal
90° phase shifter
Carrier frequency
Quadrature (90° out-of-phase)
Incidental (in-phase)

I/Q modulation/demodulation and the I/Q diagram were invented together as part of digital transmission technology. In the accompanying figure, block diagrams show the modulator at the transmitter, above, and the demodulator at the receiver, below.

The I/Q modulator takes the modulation, already separated into separate sets of information, and jogs the one designated quadrature input by 90 degrees. The I and the Q signals are summed up, keeping their phase information carefully intact, and sent along to be upconverted and combined with other channels for transmission. This signal is called a composite signal because it is made up of two distinct and independent signals. A 90 degree phase shift between them prevents either signal from interfering with the other. This relationship between I and Q signals is called orthogonality, orthogonal meaning at right angles.

Orthogonality

The right-angle relationship between two vectors means that a change of amplitude in one has no effect on the other, thus reducing internally generated distortion that is similar to cross modulation.

To understand how orthogonality helps keep two signals from interfering with one another, consider the effects of changing either the horizontal or vertical display size of your television or computer monitor. These adjustments are orthogonal to each other. Changing the size of the vertical display does not change the size of the horizontal, and vice versa. The I/Q modulator separates a vector signal into I and Q signals so they can be transmitted on the same signal carrier with little or no interference between them.

I/Q modulation

I/Q modulators translate amplitude and phase changes of a time-varying vector signal into two separate signals: in-phase (I) amplitude and quadrature (Q) amplitude. The term amplitude is used because the value can be positive (+) or negative (-).

The two orthogonal I and Q "channels" are each set to amplitudes independently by the I/Q modulator. If these amplitudes were independent analog AM signals, the diagram would look like the bird's nest of Figure 6.8 (a). If the AM waveforms were allowed only two amplitude levels, representing the two binary states, one and zero, the I/Q diagram would look more like (b), where the one/zero signal responses "collect" around four distinct locations. Each of the four collection points can be interpreted as unique values. In an I/Q modulation scheme, these are called symbols. **Symbols** are the fundamental value unit for digital transmission. More on symbols later in this chapter.

Symbols are the fundamental units of digital signals

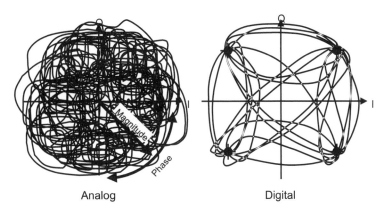

Analog   Digital

Figure 6.8. I/Q diagrams of analog and digital data.

Digital modulation and I/Q are not new

Digital modulation is the use of digital technology to apply AM and PM to a signal carrier. For over 40 years I/Q modulators have been used to send wideband terrestrial data at microwave frequencies via line-of-sight (LOS) systems in North America, radiolinks in Europe, and satellite communications worldwide. The eye diagram was the first graphical tool to be used to give system operators a look at the signal quality in both the I and the Q signals. "Closure of the eye," as you will see, was a sure sign that the QPSK signal was being damaged by noise in the transmission bandwidth.

Decision points

The information transmitted in a digitally modulated signal, that is, the symbols, are sent with a certain rhythm like the timing of an analog video's horizontal sync pulse. But in I/Q modulation, the amount of time available to collect symbol data is very short compared to the time between symbols. The symbols are read during the **decision point** time.

Figure 6.9 shows the similarities and differences between the way an analog video signal sends data, and transitions between an I/Q signal's symbols. The data in the analog signal is a large portion of the repeating data cycle, whereas, in the digital signal, the transition makes up most of the cycle time; the decision time is very short. The time scales are exaggerated for this illustration. The time between pulses in (a) are about 1000 times longer than the transitions for the digital modulation in (b). TV sync pulses trigger the receiver to accept horizontal luminance information while it blanks the picture. The sync pulse is a transition message to the TV receiver. The horizontal luminance data are transmitted between sync pulses.

In a digital modulation transmission the symbol also occurs between the transitions, but the transition is a major part of each data transmission cycle. The symbol is delivered at the decision point, a small slot in time when the receiver records the I and Q voltages. The digital receiver must know when to read the symbol. In the television receiver, the sync pulse is the receiver's trigger. In the digital receiver, the decision point is predicted by a system clock in the receiver which is synchronized by messages from the receiver. Some receivers, like set-top boxes, have this clock timing built in, and it is based upon the specified modulation standard. When a standard clock reference is not available, the receiver must be able to recover the symbol rate from the signal itself. This process is called **clock recovery**.

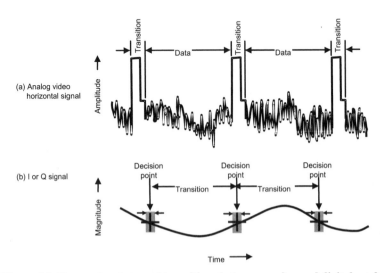

Figure 6.9. Comparing data and transitions between analog and digital modulation formats.

Jitter

Jitter can look like noise

Jitter is the short-term phase variation of the digital signal from the decision point timing.

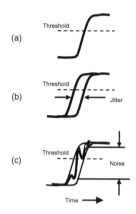

(a)

Threshold

(b)

Threshold

Jitter

(c)

Threshold

Noise

Time →

Figure 6.10. The effect of jitter on a symbol decision point.

It is usually caused by a phase instability of one of the frequency references used to convert a signal from one frequency to another. When the receiver timing is ready to read the symbol, any change in the incoming signal's timing causes data errors.

Jitter has an effect like system noise because it masks the receiver's ability to decide whether a symbol voltage has exceeded a specified threshold. To illustrate, consider the voltage transition shown Figure 6.10. The voltage transition in (a) crosses the decision threshold at one unambiguous point. In (b), jitter is a time band of uncertainty caused by the instability of the signal's time base. When applied to the signal in (a), the signal can be anywhere within the jitter band, as shown in (c). The signal amplitude is randomized just as if the signal-to-noise ratio were degraded. The result is an ambiguous threshold crossing and the chance for error in reading the symbol.

In the RF transport of signals, the instability of the oscillators used as carriers of digitally modulated signals directly affects the stability of the signal. The signal sources used to up- and down-convert signals at the head end can also contribute short-term frequency stability errors which cause jitter. Less common, but just as destructive, is the jitter from the digital signal sources themselves. Data streams that are delivered at the head end with jitter between the groups of bits cannot be distinguished between other sources of jitter, unless the data stream is read by an analyzer, such as a protocol analyzer. This type of analyzer examines the baseband data stream for the specific format. It is often difficult to track down the source of the jitter. When protocol analysis is impractical, such as when the data stream is not available, constellation and eye diagrams may help with the diagnosis.

Constellation and eye diagrams

Constellation diagrams show digital modulation

A key design and measurement tool for digital modulation is the **constellation diagram**. A constellation diagram is a plot of the points on an I/Q diagram, without showing the trails between the dots. If you picture a rectangular tube representing an I/Q diagram over time, you get a drawing like the one in Figure 6.11. The signal's vector moves forward in time down the tube passing through evenly spaced panels representing the symbol decision points. The ideal symbol point is usually indicated by a cross-hair cursor. If the modulation is high quality, the vector should go through one of the symbol point cross hairs. From the end of the tube the vector's path looks like a coiled snake, shown in (b). The points where the vector passes through the symbol decision points appear as dots on or near the cross hairs. This is shown in (c). Since an aggregate of points begins to look like a nighttime star field, this picture is called a constellation diagram.

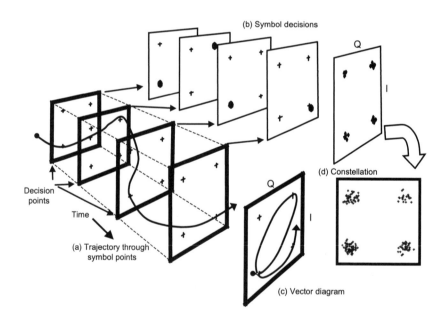

Figure 6.11. A constellation diagram built from time slices from the I/Q vector trail.

To see a composite history of the modulation, the signal is allowed to run for a number of symbol periods and is displayed as a collection of dots and vectors as shown in Figure 6.12.

Measurement instruments, such as modulation analyzers and vector signal analyzers, display constellation diagrams as part of their analysis tool set.

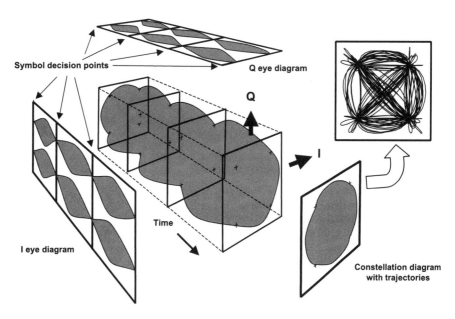

Figure 6.12. Aggregate constellation diagram formed when the vector is allowed to run for many symbols.

Eye diagrams show how the I and Q vector trails behave over time. An eye diagram is simply another view of the same I/Q time tunnel. The symbol decision points become vertical lines when viewed from the side, as a Q eye diagram, or from the top, as an I eye diagram. See Figure 6.12. The convergence of the vector trails at the symbol points is easily seen as a focusing of the trails at the symbol points.

Eye diagram	**The eye diagram shows I or Q vector paths over time. The diagram gets its name from the eye-like pattern formed by the space between the vector trails. Don't confuse the eye diagram with the term I, for in-phase. When someone says "I diagram," do they mean eye or I?**

Symbols, symbol rate, and bit rate

Here is how the bit rate, symbol definition, symbol rate, and modulation format fit together in the transmission of digital information:

1) A minimum bit rate is required by a specific application for the digital transmission. For example, a video picture needs to have a minimum bit rate for viewing.

2) A modulation format is chosen that gives the best performance for the type of transmission path, bandwidth, amount of interference, distortion, and noise. An example is QPSK being chosen for use in satellite transmissions where fading and interference is a problem.

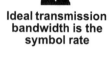

Ideal transmission bandwidth is the symbol rate

3) Since a symbol is a single transmitted value, then, in a perfect system, the transmission bandwidth would allow the symbol rate to equal the bit rate. This implies that the bit rate throughput improves with the number of bits per symbol. Since a symbol is a single "letter" in the modulation's format, the bandwidth required is equal to 1/symbol rate.

4) The practical limits of how fast the signal can be moved from symbol to symbol reduce the actual bit rate to a value less than the ideal.

Here is an explanation of this process, starting with some definitions.

A **symbol** is the lowest level piece of information that is digitally modulated onto a carrier, just as an alphabetic letter is the smallest component of a word. Symbols transmit bits. Each modulation format defines the way a symbol is to be sent and interpreted. For example, a symbol may be an exact series of bits, or it may be the difference between a sequence of bits. A sequential series of bits is called a **bit stream**. The bit stream 10110101 may be defined as a symbol in one digital modulation format. In another format the change between 1001 and 1101 may define a symbol. The exact way symbols are interpreted by specific formats is beyond the scope of this book, but it is important to know how I/Q modulation forms unique symbol sets.

Symbol and symbol rate

Symbols are the letters in the language of digital modulation. A symbol is assigned the duty of transmitting one or more digital bits of information. The symbol rate is the number of symbols sent per second. The reciprocal of the symbol rate is the approximate bandwidth necessary to transmit the signal.

Different modulation formats require different collection of symbols, just as the English and Russian alphabets have a different number of unique symbols to allow the formation of the words required for communication. Modulation formats have varying sets of symbols depending upon speed and bandwidth requirements.

Baud is not bit rate

Often bit rate and baud are use interchangeably. This is incorrect even though both are used to indicate the rate of information flow. Bit rate is the number of binary digits, 1's and 0's, that are transported through the system every second. Baud is the number of signal changes that can occur in a transport system every second. The signal changes may or may not relate to the number of 1's and 0's depending upon the signal's encoding formats.

Each dot or crosshair on the constellation diagram of an I/Q modulation format represents a unique symbol. Each symbol is a unique binary number having the same number of digits. The simple I/Q diagram in Figure 6.13 illustrates. There are four symbols in the modulation format. Using the formula in Table 3.1 on page 43, the number of bits required for four symbols is 2. How the values are assigned to each symbol depends upon the format definition; the assignments are optimized to reduce ambiguity between symbols while making vector transitions between symbol points efficient. The symbol's binary values are assigned to the four points in the figure, ranging from 00 (zero) to 11 (three).

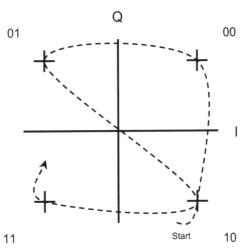

Figure 6.13. Four symbol positions in an I/Q diagram.

To send a data stream, the modulated carrier changes it phase and magnitude to pass through the symbol points in a sequence given by the baseband data. For example, to transmit the data stream 1000011011 on the I/Q diagram of Figure 6.13, the vector moves through the symbols, as the dashed line, starting in the lower-right corner at 10, then moving sequentially through 00, 01, 10, 11. Constellation displays in test equipment can usually be set to show this trail.

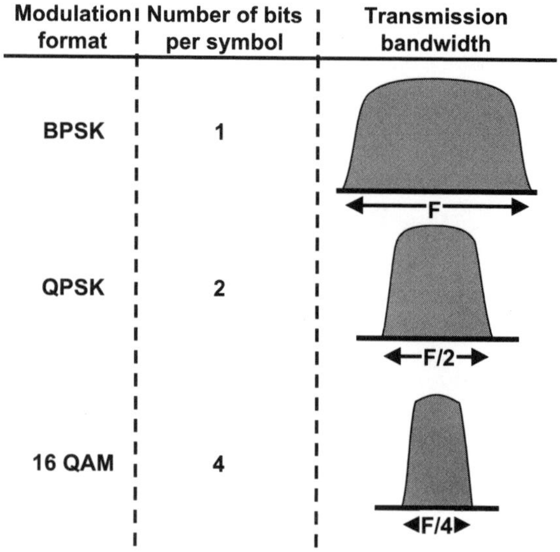

Modulation format	Number of bits per symbol	Transmission bandwidth
BPSK	1	F
QPSK	2	F/2
16 QAM	4	F/4

Figure 6.14. Relating modulation format, bits per symbol and transmission bandwidth for the same data rate.

Because a symbol is the smallest unit transmitted in the modulation message, just as the bit is the smallest unit in a binary number, it makes sense to refer to symbol rate in Hz. **Symbol rate** is the rate at which symbols are sent in units of symbols per second. Thus, symbol rate in Hz is the frequency spectrum required by a modulated signal to transmit all its symbols intact. If the symbols could be set as fast as possible, the symbol rate would equal the bit rate and therefore, its **transmission bandwidth**. But symbols can only be sent so fast without causing extreme interference with neighboring channels, so they have to be contained with filtering, which requires slowing the symbol rate, and therefore, the bit rate. See "Modulator filter effects" on page 95.

For the same transmission data rate different digital formats require different transmission bandwidths. Figure 6.13 shows three formats: binary phase shift keying (BPSK), QPSK, and 16-QAM. For each to send a signal with the same data rate, the transmission bandwidth must be inversely proportional to the number of bits in the format's symbol. The more bits per symbol, the narrower the transmission bandwidth. 16-QAM can fit the same data into half the bandwidth of QPSK, for example.

Usually a modulation format is specified by its symbol rate. Here is the simple formula for determining if the format supports the bit rate required.

$$\text{bit rate in bps} = \text{symbol rate} \times \text{bits per symbol}$$

This formula can be turned over.

$$\text{symbol rate in Hz} = \text{bits rate}/(\text{bits per symbol}).$$

For example, if a composite digital television channel needs to be sent real-time at 135 Mbps, the bit rate is 135 Mbps/2, or 67.5 Msymbols/sec, since there are two bits in each symbol. The digital format of Figure 6.13 would require a symbol rate of 135 Mbps/2, or 67.5 MHz to transmit, using a great deal of the cable system's available bandwidth! A format with 256 symbols available provides 8 bits of value ranges and the 135 Mbps bit stream can be transmitted in just 16.875 MHz (135 Mbps/8).

Modulation format	Bits per symbol	Bit rate	Typical symbol rate
QPSK	2	256 kbps	128 kHz
QPSK	2	2 Mbps	1 MHz
QPSK	2	10 Mbps	5 MHz
8-VSB (U.S.)	3	15 Mbps	5 MHz
16-VSB (U.S.)	4	20 Mbps	5 MHz
16-QAM	4	20 Mbps	5 MHz
32-QAM	5	25 Mbps	5 MHz
64-QAM	6	30 Mbps	5 MHz
256-QAM	8	40 Mbps	5 MHz

Table 6.1. Symbol rates based upon number of bits per symbol and bit rate.

Table 6.1 shows the best bit and symbol rates for several modulation formats using the formulas above. Transmission bandwidth needs to be greater than the required symbol rate because the symbol transitions need to be smoothed by filtering to keep their energy from splattering outside the transmission channel and interfering with other signals. For example, the same bit rate can be transmitted in a 36-MHz bandwidth QPSK formatted signal and a 6-MHz bandwidth 64-QAM signal. QPSK has 2 bits per symbol, and 64-QAM has 6 bits per symbol. If symbol rate could be made into transmission bandwidth, you would expect a 6/2, or three-fold increase in bandwidth due to the number of bits per symbol. But the actual bandwidth requirements are 36/6, or six-fold due to the practical speed at which a symbol can be sent, that is, actual symbol rate.

Because there are different number of bits per symbol, the same 5-MHz bandwidth can be used to transmit a wide range of bit rates. The system designer chooses the format appropriate to the application. If the information to be transmitted requires high bit rates, such as a complex digital video signal, then a modulation using a higher number of bits per symbol can be used as long as the transmission path is free from noise and other impairments. Increasing bit rate within the same transmission bandwidth, implying that the symbols have more bits, has the drawback that the signal is more susceptible to bit errors. For a return path in the 5 to 42 MHz cable system, where the interference and noise are likely to be high, QPSK with its clearly defined symbols is used to keep BER low.

More bits per symbol requires transport with less noise

Real-time

Real-time means that delay between an event, such as a television frame reception at the head end, and its use at the subscriber's TV does not disrupt the basic functionality or the primary use of the system, in this case to enjoy television. From the subscriber point of view, longer delays than the near-speed-of-light cable transmission are still considered real-time. The delay doesn't affect the viewers ability to enjoy the material. In a signal process, real-time means that there is little or no delay between the input and output. However, if the subscriber were using the cable for interactive video, such as a game, or voice/picture communications, a delay as little as 0.3 seconds is objectionable.

This seemingly small delay, the 0.3 second wait for the speed of transmission over a satellite up-down link, is very annoying when applied to a two-way phone conversation. Delays even this short cause overlapping dialog, misunderstandings, and awkward silences.

The transmission frequency is important in the computation of the cable's capacity. For **real-time** transmissions, such as video, a modulation format and bit rate are chosen for the transmission media which keeps the data flowing with acceptable quality in the smallest transmission bandwidth. Signal quality parameters and measurements are the subjects of the next chapter.

Modulation format choice	**To transmit higher bit rates in the same bandwidth requires trading off the vulnerability of the signal to interference and transmission impairments. Lower bit rates can be transmitted with very robust formats, such as QPSK.**

Modulator filter effects

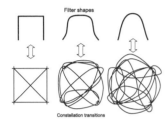

Filter shapes

Constellation transitions

The modulator's filter reduces the frequency response of the amplitude/phase changes as the carrier moves from symbol to symbol. The illustration shows the effect in three successive pictures, left to right, of a QPSK signal as fewer high frequencies are allowed through the processor. Another way to think of this effect is to look back at the reconstruction of a square wave from its harmonics in Figure 2.10 on page 26. As the higher-frequency harmonics of the square wave are removed, the wave becomes more rounded, with less sharp corners. This smoother transition take more time, so symbol rates, and therefore, the transmission bandwidth, are reduced.

The reduction of the higher-frequency components of the vector signal allows its energy to be confined to a smaller bandwidth. The filter shape also affects the frequency band edges, providing the characteristic rounded top to the signal in the frequency domain. The illustration at the right, which shows the time and frequency domain displays of different-shaped signals, is called a transformation diagram. It demonstrates that when pulses of RF energy in time are smoothed, their corresponding frequency domain signals are also smoothed, as shown in this next drawing.

The top pulse, the square wave on the left, is a main lobe signal with two smaller lobes to each side in the frequency domain on the right. Since most of the energy is in the main lobe, the side lobes are wasted, and could interfere with adjacent channels. To reduce the side lobes, the pulse shape in time is changed, tilting the leading and trailing edges of the time pulse inward, and the frequency side lobes begin to diminish. When a smooth-curve filter is applied to the pulse, at the bottom, the frequency response is almost the same shape. Now all the energy of the signal can be contained in the main lobe, adjacent channel interference is reduced, and little channel bandwidth is wasted.

Time ⟶   Frequency ⟶

Filtering a digitally modulated signal

The ideal transmission rates cited for digital modulation formats are never achieved in actual systems because the ideal rates assume that the signals are not filtered. Yet filtering the transmission bandwidth of a digitally modulated signal is required to keep the signal in its channel. See the box below. The filtering is applied at the digital modulator or as a digital process shared between transmitter and receiver as part of adaptive equalization. From a maintenance standpoint, the good news is that digital filtering is less apt to go bad. If it does, the repair mode is to find the offending modulator and replace it with a good one.

With an understanding of the way digital data are modulated onto a carrier, you are equipped to learn about the way digital signal quality is measured in Digital Signal Quality, Chapter 8 beginning on page 127. But first, the following chapter, Error Correction, Equalization, and Compression, describes the techniques that help make a digital signal a managable size and robust while being carried by the digital modulation carrier.

Summary

The complexity of modulation formats continues to increase in response to communication needs. Most, if not all, formats invented since 1990 are digital, even though digital techniques have been used for over 40 years, with their beginnings in terrestrial and satellite microwave data links.

Digital modulation prepares the encoded and layered digital signal for transport. Although combinations of AM and angular modulation (FM and PM) are used, there are significant differences between the way signals are presented for analysis. In-phase and quadrature signals, I/Q, provide the modulation and the view required for efficient testing. There are many digital modulation formats, but the ones of interest for cable systems are QPSK, QAM, VSB, OFDM, COFDM, listed roughly in order of complexity. FDM, TDMA, and CDMA are multiplexing techniques, not modulation formats. These multiplexing techniques are usually associated with specific modulation formats, especially in cellular radio applications.

Analog signals show their performance and content in the frequency and time domains. Generally, digital signals look the same whether they are carrying information or not because the signal is layered with coding in addition to its modulation encryption.

Orthogonality means that a carrier may be modulated with two independent information signals provided the vectors produced by the modulation are always shifted 90 degrees from one another. The vector diagram is a display of the modulation amplitude and phase of the modulation vector as viewed from the carrier. I/Q modulators take two independent signals and modulate them onto a carrier orthogonally.

The information in a digital signal, the symbols, are transmitted at a constant rate. Each symbol is assigned a specific I/Q value. The symbols are defined at a specific time, called a symbol decision point. Knowing the decision points, a receiver or test instrument can read the symbols and decode to get the transmitted bit stream.

Since the modulation is in-phase and quadrature, it is convenient to view the modulation in the I and Q planes using eye diagrams and constellation diagrams. The constellation diagram shows the symbol points in an I/Q graph collected for a specified number of symbols. The eye diagram shows the path between symbols either along the I or Q plane. A number of subjective measures of modulation and signal quality can be made from constellation and eye diagrams.

The bandwidth of a digital signal is roughly given by the bit stream rate divided by number of bits per symbol. The higher the number of bits in a symbol, the less bandwidth required to transmit a given bit stream rate. When more bits are packed into a symbol, they are harder to distinguish at demodulation, causing more errors.

Filtering of the modulated signal keeps the signal within its channel bandwidth, while using filter shapes, such as a raised cosine, to keep symbol vectors from interfering with each other. Jitter is the short-term phase variation of the digital signal from the ideal symbol decision points in time.

Questions for review

1. A vector diagram shows changes to a carrier due to modulation. If the vector is moving slowly clockwise around the origin with a constant radius: (select all that are true)
   a) there is no amplitude modulation
   b) the carrier is phase modulated with a CW signal
   c) the frequency of the carrier is slowly increasing, but there is no other modulation
   d) amplitude modulation is causing a frequency drift
   e) none of these

2. A digital signal is separated into in-phase (I) and quadrature (Q) signals for transmission
   a) to maintain high transmission quality because I and Q signals do not interfere with one another
   b) because the modulators for digital signals create I and Q signals
   c) all of the above

3. A symbol is to a digital modulation signal as a letter is to an alphabet. True or False?

4. A constellation diagram shows (check all that are true)
    a) dots representing symbols at the symbol decision points
    b) dots representing symbols between the decision points
    c) information that enables signal and system troubleshooting
    d) all of the above

5. A digital modulation format has 16 bits per symbol. What is the bandwidth necessary to transmit a 256 kbps digital signal?
    a) 16 MHz
    b) 1 MHz
    c) 256 kHz
    d) 16 kHz
    e) none of these

Selected Bibliography

1. *Digital Radio Theory and Measurements, an Introduction to Digital Radio Principles, Practical Problems and Measurements*, Application Note 355A, publication number 5091-4777E, Hewlett-Packard Company, 1992.

2. *Interval*, the SCTE Newsletter, Society of Cable Telecommunications Engineers, Inc. *DigiPoints*, Exton PA, June 1997.

3. Helen Chen, "Testing Digital Video: A Look at Measuring Power and Interference," *Communications Technology* magazine, Phillips Business Information Inc., Rockville MD, May 1995.

4. Helen Chen, "Testing Digital Video: New Measures of Signal Quality," *Communications Technology* magazine, Phillips Business Information Inc., Rockville MD, June 1995.

5. Dr. Kamilo Feher and the engineers of Hewlett-Packard Ltd., *Telecommunications Measurements, Analysis, and Instrumentation*, ISBN 0-13-902404-2 025, Prentice-Hall, Inc., Englewood Cliffs, New Jersey, 1987.

6. Deiter Scherer, "Measurement Tools for Digital Video Transmission," Hewlett-Packard Video Communications Division, *IEEE Transactions on Broadcasting*, Volume 39, No. 4, December 1993.

7. David R. Smith, *Digital Transmission Systems*, ISBN 0-442-00917-8, 2nd Edition, Van Nostrand Reinhold, New York, 1992.

8. Wayne Smith, paper for the HP 89410A/41A Vector Signal Analyzer field training event, Hewlett-Packard Company, June 1996.

9. John Watkinson, *The Art of Digital Video*, ISBN 0-240-52369 X, 2nd Edition, Focal Press, Oxford, 1995.

10. Helen Wright, *Digital Modulation in Communications Systems - an Introduction*, Application Note No. 1298, publication number 5965-7160E, Hewlett-Packard Company, 1997.

7

Error Correction, Equalization, and Compression

A digital signal's robustness, quality, and noise-like nature are provided by the processes that build it. This chapter helps you understand enough about these processes to understand the nature of the signal, without burdening you with the details of how they work in detail.

What you will learn

- Error correction, equalization, and compression uses
- Why do digital signals fail?
- What is the cliff, or waterfall, effect?
- What are transport impairments?
- How does a digital signal fix itself?
- What measurement parameters check the effectiveness of error correction and equalization?
- How does compression affect transmission bandwidth?
- What are the measures of delivered quality?

The keys to signal quality and efficiency

Standards like DVB-C are coming

You can no more tell digital signal quality from its performance at the receiver than you can tell a book by its cover. A digital signal keeps its data safe and your view hidden from the adversities of transport between the head end and the subscriber by using a number of signal encoding processes. For most video applications today, error correction, equalization, and compression are required by the format, modulation and distribution standards. It helps that these processes are readily available through inexpensive computer power. The European DVB-C standard is a ground-breaking example. Standards are now being written whose effect will be to improve system performance and test cost effectiveness for most world-wide cable systems. Undoubtedly these standards will include some form of error correction, equalization and compression. Figure 7.1 shows a map of these processes as factors in digital signal quality. A reasonable understanding of these techniques is necessary to understand the composition and testing of digital signal quality.

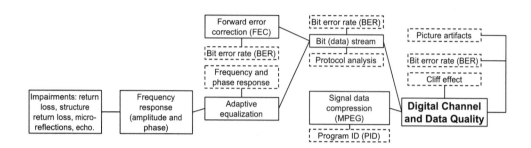

Figure 7.1. A map of the influences of encoding on digital signal quality.

Figure 7.2 shows the placement of these processes in the signal layering. Each encoding process adds to the signal's efficiency and robustness.

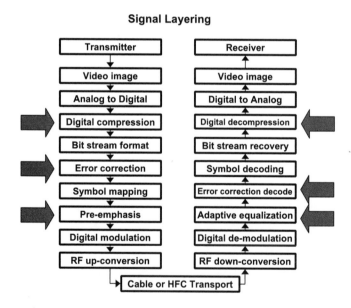

Figure 7.2. The positions of compression, error correction, and equalization in the signal layers.

How an Analog Wave Becomes Digital Data, Chapter 3 beginning on page 33, says that the digital signal is inscrutable. Compression, error correction, equalization, and modulation are partly responsible.

A digital signal cannot be quickly inspected in the RF domain as is possible with analog TV signals. It is important to open up the digital signal processes to evaluate signal quality, compare performance to the delivered quality, and determine how close the signal is to the edge of the cliff. To this end, the rest of this chapter describes error correction, equalization, and compression. Each topic contains

- the encoding process purpose,
- the system impairments corrected,
- the effect on the measure of system quality,
- and hints on how to troubleshoot when things go wrong.

The causes of errors

Impairments come in many forms

Distortion, interference, noise, hum, poor frequency response, and reflections all degrade analog and digital signals. The robust nature of digital signals allow them to maintain high quality over a great variety of impairments, but can conceal problems in system performance such as

- **Attenuation distortion and frequency response distortion:** If the frequency response of the signal changes from the transmitter's output due to attenuation slope or filtering edge effects, such as with a suckout, the receiver cannot adequately reconstruct the signal.
- **Envelope delay distortion, group delay:** Different frequency waves travel though the medium at different speeds. For a complex signal to maintain its frequency response and time waveform shape, all the frequency components must arrive at the receiver in an orderly fashion, otherwise the phase changes cause distortion similar to that of the attenuation distortion.
- **Signal-to-noise ratio:** Noise in a signal's channel passband can cause the symbol points to become blurred enough that the modulator cannot read them.
- **Harmonic distortion:** The signals generated by the mixing of many channels, such as the familiar CSO/CTB, causes composite distortion which can fall in the digital channel, scrambling or masking the symbols just like excessive noise.
- **Jitter (phase jitter):** Digital signals need precise system timing coordinated with the receiver to read the incoming symbols correctly. When the timing between groups of bits, or between the transmitter and receiver clocks is off, symbols are misread.

- **Impulse noise:** A short pulse of unwanted energy on the transmission line can temporarily remove a great number of symbols.
- **Echo, or microreflection:** The reflection of a portion of the transmitted signal toward the transmitter, caused by mismatches in impedance within the transmission path, sets up a reflection that distorts the transmitted signal.

All of these impairments cause digital signal errors. Some amounts of attenuation distortion, group delay distortion, and echo can be corrected by adaptive equalization. Error correction encoding and decoding tries to fix the damage to the data stream caused by any impairment.

Error correction

Error correction tries to keep the signal sailing

The sole purpose of error correction is to keep the signal at top quality up through any type of impairment. But digital signals fail with a crash when impairments overwhelm signal, equalization, and error correction. Then the receiver no longer can present the data and the picture goes black, or the application fails. This is called the **cliff effect**. The cliff effect is quick, unlike the effect of increasing impairment of an analog signal which tends to degrade in proportion to the adverse conditions.

Error correction in everyday life

Personal error correction is when you deduce the meaning of a badly misspelled word, or one with an S or N mirrored, by taking into account the word's source or phonetic soundings. In other words, you draw on a great deal of knowledge about language, symbols, and environment to make the error correction. But a data package sent over miles from one machine to another does not have the luxury of a receiver intelligent enough to guess the content of a garbled pile of bits. Error correction is the ability of a data stream to self-correct, without external data being added at the receiver. It is so important to the transmission of data that fully one third of the bits in an audio CD are for error correction.

Forward error correction, or **FEC**, simply means that the error correction information is carried forward to the receiver. No additional data are required at the receiver to restore the lost bits. A more data-oriented definition is from Newton in the bibliography at the end of this chapter. It states that FEC is "a system of data transmission in which redundant bits generated at the transmitted end are used at the receiving terminal to detect, locate and correct any transmission errors before delivery to the local data communications link."

Forward error correction

Forward error correction, or FEC, is a digital transmission system that sends redundant information along with the payload so that the receiver can repair the damage and eliminate the need for retransmission. The extra data reduces the size of the payload. In real-time applications, such as video, retransmission is too late to fix a broken picture.

Bit, packet and frame error rates

FEC masks true BER

Bit error rate (BER) is an objective way to measure the quality of digital transmission. The measure of BER is being replaced by the measure of other quantities more commonly used in the transmission of digital data, namely packets for multiplexed streams, and frames for video information. These terms are also used in the cellular wireless transmission formats. But for the purposes of learning how digital signal quality is maintained and corrected, the term BER is used.

Forward error correction improves the bit error rate of the digital information between the encoding at the head end, or a digitizing center, and the decoding at the subscriber's set-top box. BER, covered in detail in the following chapter, can improve a signal from 1 bit lost in 10,000, to just 1 bit lost in 100,000,000,000! But in doing its job, FEC hides the true degradation of the signal quality from the service technician, so that distance to the cliff, called **margin**, is not known. The test points for measuring true BER must be chosen carefully, and a specific test plan is required. In some types of FEC implementations, error statistics are provided at the receiver so you can get a quick, subjective look at how close to the edge of the cliff your equipment is running.

Margin, how close to the cliff?

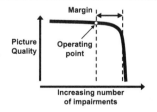

Margin is a term used to describe how close your digital signal is to becoming unusable, or how close to the cliff edge the system is operating. Digital Signal Quality, Chapter 8 beginning on page 127, discusses the measurement of margin.

In other cases you may have to rely on a process for the direct measurement of the digital modulation quality to see the effects of impairments being concealed by equalization and error correction. Unfortunately, this usually requires taking the system out of service by replacing the program material with specially coded digital test signals. However, eye and constellation diagrams observe the effects of impairments on signal modulation without disrupting service. Alternatively, a non-intrusive BER test can be done using the error information in the receiver's forward error correction chip. This technique is discussed later in this chapter.

Reed-Solomon error correction coding

Reed-Solomon coding does FEC

Transmitted information, by definition, cannot be predicted by the receiver, and the receiver cannot deduce a correction. It must have access to additional duplicate versions of the payload. This redundant data are sent encoded differently from, but along with, the payload. Any contrast in the data patterns of the two forms of data initiates decoding of the redundant data, which is used to replace the damaged payload. Otherwise, the redundant data are discarded.

Error correction is trivial, but...

If you know that a bit of value zero is wrong, changing it is trivial. Just make it a one. The key to error correction is determining **which** values are in error. In error correction the "where" is determined by the codeword which contains information enough to correct a predictable number of corrupted bits. The codeword is generated as an overview of many bits, transmitted over a greater time than any one data bit. The codeword contains the same data based upon the payload information, but coded in a different way, and over a different time period. This reduces the chances that the codeword will also be in error.

The ratio between redundant information and payload is dictated by the requirements of the signal and system. The factors that determine this ratio are the desired minimum error rate, the type of payload, and maximum data rate. The higher the ratio of redundant data to payload, the better the chances for fixing the received signal and lowering the BER, at the expense of bit rate and/or the size of the payload.

Reed-Solomon ideal for bursts, and customized for specific system

Reed-Solomon forward error correction coding is named for mathematicians Irving Reed and Gustave Solomon, who presented their work in the early 1960s. Three attributes make Reed-Solomon coding well adapted for video and data:

The coding

1) is inherently **burst correcting** because it works on the multibit symbols rather than individual bits. Code that is interleaved for

multiplexing data streams benefits since large groups of bytes are less likely to be interfered with by bursts of noise or interference unless the interference lasts longer than the interleave cycle time.

2) can be **customized** for the desired level of redundancy at the design stage, that is, the number of correctable bursts can be set. The number of bursts which are correctable can be chosen at the design stage by the amount of redundancy required. For example, in European 64-QAM digital video, the selected level of Reed-Solomon coding is set such that it adds 16 parity bytes to the 188 bytes of data, so that 204 bytes of data are transmitted. The extra data allow up to 8 byte errors in the primary 188 byte data stream to be corrected, if necessary.

3) **operates very close to the theoretical limit** for error correcting. When selecting the level of error correction for a specific application, the level must be appropriate for the probability of errors to be encountered. An inadequate error correction system is actually worse than not having any correction. Error correction works by trading off probabilities. Error-free performance with a certain error rate is achieved at the expense of performance at higher error rates, so if the designed bit rate is exceeded, FEC itself can crash the signal. Fortunately, in most video delivery systems, the data rate is fixed or has some practical upper limit.

What FEC statistics can tell

FEC statistics, available at the receiver, show how hard the error correction mechanisms are working. The harder they work, the smaller the operating margin, that is, closer to the cliff. Operation parameters for FEC in the DVB-C data stream are primarily a count of the total cumulative error bytes and packets counted and a ratio number of the bad bytes and packets to the good ones. One such measurement is displayed in the screen of a DVB-C, QAM analyzer, shown in Figure 7.3. The error rates are compared to the design standards to provide an on-going margin for the test point.

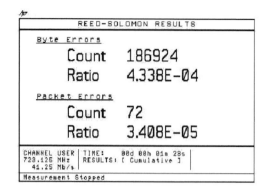

Figure 7.3. Reed-Solomon byte and packet error count displayed on a QAM analyzer.

This measurement checks the data integrity by looking at the data stream for correct decoding in real-time, and without interfering with service. The decoder can also count and record the number of bit errors that have been corrected. The FEC chip can supply an accurate BER without injecting a known bit stream. This non-intrusive BER depends on the power of the FEC to detect bit errors. Different FEC codes have different error detection sensitivities depending upon their coding parameters. If the BER exceeds the error correction power of the FEC, the error statistics do not accurately reflect the true BER. In other words, if it is probable that every packet in the data stream has at least one error, the BER is underestimated.

Another parameter that allows real-time quality monitoring is the measure of the program identification in the DVB-C MPEG data stream. This is discussed in "Trouble from compression" on page 121.

Adaptive equalization removes linear distortion

Adaptive equalization removes distortion effects

Adaptive equalization, or AE, improves end-to-end BER by removing linear distortion effects from signal caused by certain transmission impairments. On a macro scale, equalization is used in cable television amplifier chains to compensate for the slope introduced by the frequency-dependent losses in the coaxial cable. This equalization is done with equalizer circuits in the cable amplifier diplexers. Adaptive equalization is compensation of the signals themselves for distortion created by all the components in the cable system. The equalization occurs real-time, that is, as the signal is being impaired.

The signal conditioning, called pre-emphasis, compensates the baseband signal just prior to its modulation, as shown in Figure 7.2 on page 100. At the receiver, the adaptive restoration occurs right after the signal is demodulated.

Linear transmission impairments are frequency response errors, such as ripple and tilt, reflection distortion, non-flat group delay, and phase distortion over the bandwidth of the signal. **Group delay** is the time it takes a signal to move through the system. Signals at different frequencies travel at different speeds, arriving at the receiver at different times. If the time spread is too great, the waveform of the received signal is distorted. See the "Group delay" explanation below. There are many sources of linear distortion in a system:

- bandpass filters
- improper cabling terminations
- component mismatch
- signal combiners
- poor connections

AE helps you measure other system problems

As with the FEC process, the real-time correction of system impairments can mask transportation impairment problems. But adaptive equalization allows the receiver test equipment to look at signals otherwise too distorted for measurement. Figure 7.4 shows an example of a 16-QAM signal impaired by linear distortion, on the left, and the recovered signal on the right.

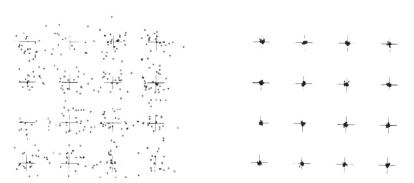

Figure 7.4. 16-QAM signal as received (left), and as processed by an equalization filter (right).

On the right, is the same signal after being processed through an appropriate level of equalization. The convergence of the symbol points around the symbol targets shows the removal of the excessive linear distortion affecting the signal.

Group delay

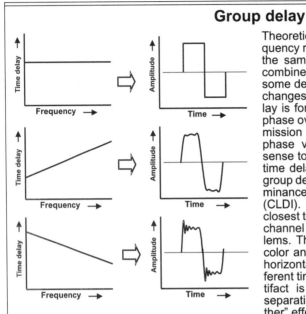

Theoretically, all frequencies within a cable's frequency response should arrive at the receiver at the same time. But signals are processed by combiners, filters, and active devices that put some delay in the signals' progress. These time changes are known as **group delay**. Group delay is formally defined as the rate of change of phase over the frequency response of the transmission medium, and it is often plotted using phase versus frequency. But it makes more sense to think of phase changes in terms of the time delays they cause. One quick indicator of group delay is the measurement of composite luminance to chrominance delay inequality (CLDI). In NTSC systems, the channel usually closest to the return spectrum in NTSC systems, channel 2, is usually the first to show CLDI problems. These are indicated by the arrival of the color and black-and-white signals for the same horizontal line of an analog video signal at a different times at the receiver. The CLDI picture artifact is a color smear at the vertical line's separating color bars, the so-called "pink panther" effect.

But group delay can have a much more destructive effect on digitally modulated signals. In a system that is required to send digital pulses, like those shown in the top drawing at the left, arrival of all the square wave's odd harmonics recreates the square wave. See Figure 2.11 on page 27 for a discussion of this concept.

In the second set of drawings, time delays cause the lower frequency odd harmonics to arrive first, causing ripple in the square wave. This ripple is smooth, so the integrity of the pulse's shape is not too distorted. When the time delays are reversed, as shown in the lowest drawings, the high-frequency odd harmonics arrive early, and gang up to cause the leading edge of the pulse wave to overshoot. The overshoot response is much more difficult for the demodulator/receiver to interpret. The result is an increase data stream errors.

Usually there is a tolerance to group delay of a specified duration in the receiver so that the signal is not distorted beyond recognition. For example, a telephone voice channel may have delay of up to 2 ms at the high and low ends of its frequency range compared to the center of the band without causing significant distortion to the listener. But a high-frequency pulse waveform used to transmit a digital message subjected to group delay can become so distorted that it cannot be interpreted by the receiver.

Limitations to adaptive equalization

AE cannot remove non-linear distortion

Distortion caused by amplifiers in saturation, such as intermodulation distortion (third order, CTB), harmonic distortion (second order, CSO), spurious responses (interference), adjacent channel interference, and noise cannot be reduced by adaptive equalization. These non-linear distortion products are the "remainders" after the equalization stage. Non-linear distortion are the changes to signals acted upon by non-linear active devices such as amplifiers in saturation and active frequency converters (mixers). Noise is not reduced by adaptive equalization because it masks the legitimate signals with random energy that cannot be filtered or predicted.

Figure 7.5. Adaptive equalization used to reveal non-linear spurious interference.

AE may make a measurement possible

For measurements, this limitation may be an advantage if it provides a way for test equipment to view non-linear transportation problems. This is illustrated in Figure 7.5. The constellation on the left is too distorted to make any measurement. After using adaptive equalization in the measurement receiver itself, an interfering spurious signal is revealed, shown in the constellation on the right. The spur is the remaining non-linear impairment. Format-specific analyzers, such as the DVB-C QAM analyzer, can measure the amount of distortion of the spur accurately.

How adaptive equalization works

Adaptive equalization applies a filter to a signal at the receiver to remove or compensate for the effects of linear distortion. The filter is defined in the frequency domain by frequency response parameters such as gain, phase or group delay. Alternatively, it can be defined in the time domain by its impulse response.

What's your delay?

A signal in space traveling from satellite to satellite moves almost at the speed of light, 186,000 miles per second (300,000 km per second). The speed of a signal in the cable or fiber is slower than the speed of light because of its resistance to the density of the cable's dielectric material. A measure of the speed of a signal through a cable is called the propagation factor. A propagation of 0.9 means that signals within the cable's specified frequency range can travel at 9/10 the speed of light through the cable. A signal in cable alone can go from the head end to a subscriber 25 miles away in just $25/(0.9 \times 186,000) = 149$ µsec (149 micro seconds, or 149×10^{-6} seconds). In fiber, the propagation is slower than cable, with a typical propagation of 0.67. The signal in the example would arrive in $25/(0.67 \times 186,000) = 201$ µsec.

Far longer delays are caused by the processing of digital signals being routed, cleaned up, and reconverted throughout the system. This delay, sometimes called **latency**, can be hundreds of times longer.

To illustrate, take a channel with a notch in the frequency domain as shown in Figure 7.6. An adaptive equalizer responds to this flatness error by constructing a filter to compensate for the problem. The filter response, shown in (b), is shaped to cancel out the frequency response of the system. The rise in the receiver passband compensates the dip in the transmission path passband. The result, in (d), is a flat response. This process uses a time domain impulse response, in (c), that dynamically creates the frequency response, in (b), required to correct the flatness problem. This illustrates the simplest of equalization, called feedforward equalization. Most systems today require much more extensive equalization, such as decision feedback equalization. Decision feedback equalization is required to keep complex digital modulation signals, such as 32-QAM and higher order formats, from being affected by microreflections. See "Finding microreflections" on page 114.

Group delay, equalizers, and diplexers

The components that separate forward and return signals in a cable system cause phase shifts at frequencies near the rolloff edges of their bands. These areas are marked by the gray rectangles in the drawing. This is unavoidable because of the way filters operate. Channels near the edge of either the forward or return bands are subject to more group delay than other channels in the system, so avoid assigning complex digital signals to frequencies near these band edges.

How does the adaptive equalizer know what shape response to create? Usually the signal itself contains a bit stream that is known by the receiver. Such a bit stream is called a **training sequence**, because it is used to compare the receiver's knowledge of what should be sent to what has actually been sent by the transmitter. The training sequence is used by the adaptive equalizer to construct the correction filter. When no training sequence time domain input response is available, the AE must take the time to iterate to a solution.

Adaptive equalization use and disuse	In almost every communications link, adaptive equalization of some sort is used. For example, in wireless applications, where linear multipath distortion is a substantial and fast-reacting problem, adaptive equalization is a required process. Two places where equalization are not used are satellite to earth station communications where there is no multipath and very few other linear distortions. In CDMA applications, adaptive equalization is of little value because the coding gain is so high that linear errors are negligible.

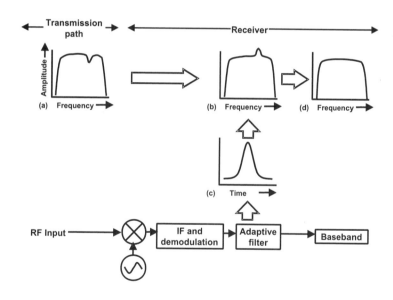

Figure 7.6. The process of adaptive equalization.

The receiver's equalizer sees system's linear response

View system response from the equalizer

The adaptive equalizer in the receiver can provide system response information for the analysis of linear transmission distortions. A receiver, or measurement instrument acting as a receiver, has the real-time system linear response stored as the frequency and time/impulse response in its adaptive equalizer's memory. This means that the linear distortion characteristics of the system can be viewed directly. This can be useful, even in systems which do not incorporate equalizers in the receivers. The adaptive equalizer allows the system transfer function to be determined without special test signals, and without direct access to the signal source. An example of the output possibilities from a vector signal analyzer used as a receiver for a digital cellular channel is shown in Figure 7.7.

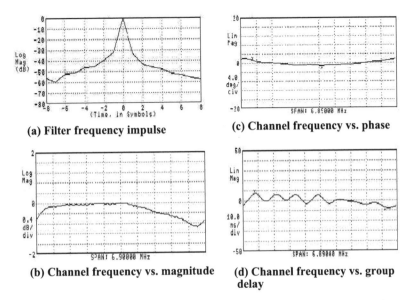

(a) Filter frequency impulse

(c) Channel frequency vs. phase

(b) Channel frequency vs. magnitude

(d) Channel frequency vs. group delay

Figure 7.7. Adaptive equalizer responses to linear path transmission distortions for a wireless signal cellular radio signal.

The impulse response of the filter is shown in (a), along with the frequency responses of magnitude, (b), phase, (c), and group delay (d). The impulse response is used to diagnose reflections from components along the cable path, multipath reflections coming from a signal originating at an antenna feed at the head end, or from an impedance mismatch at the junctions of the components in the cable plant such as the taps or trunk amplifiers. Channel frequency magnitude response shows the system amplitude alignment over a single channel. Adjacent channel responses may help see larger-scale response problems, although RF cable television sweep systems are better suited for system-wide frequency response measurements. The phase and group delay responses show the deviation of phase and signal delay across the channel, respectively, and, for cabled systems, may be indications of faulty passive devices such as combiners. These outputs from the adaptive equalizer are the inverse frequency response of the media through which the signal has passed.

A signal analyzer dedicated to analyzing specific video formats needs to have the same functional blocks as the subscriber's receiver, the set-top box. An example of the frequency and impulse response shown in a QAM analyzer for the DVB-C standard is shown in Figure 7.8.

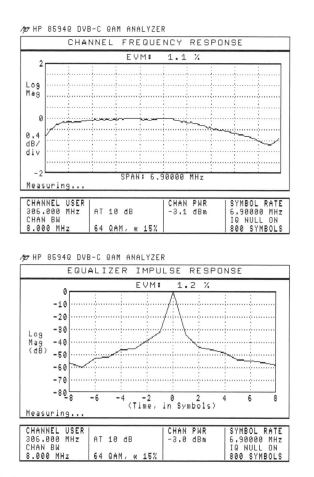

Figure 7.8. QAM analyzer's real-time adaptive equalizer responses.

The QAM equalization is presented on the display real-time. Here are the system characteristics you can view:

- level of linear distortions being added by the network
- power suckouts across the channel
- microreflections in the transmission medium

In the future it may be possible to characterize the set-top box demodulator when access to its internal equalizer is available.

Finding microreflections

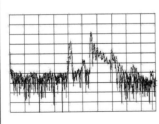

Advanced adaptive equalization, such as decision feedback equalizers, reduce the effect of microreflections on signals. But there are times when you may be required to hunt down microreflection sources. Since the equalizer measures the frequency response of the transmission cable using the incoming signals, it may be used to find microreflections in and around the set-top box, such as problems associated with the subscriber's house wiring. Here is an example using a DVB-C QAM analyzer. Real-time equalization shows the response of the analyzer's filter. The display is in the time domain, and for convenience of measurement, scaled in terms of symbols. This impulse response identifies echoes and distance to the source of microreflections from discontinuities and bad connections.

To identify the distance to the microreflection, the impulse to the right, a few calculations are necessary. For a baud rate of 6.9 Msymbols/sec, one symbol is 0.144 microsec. This large reflection at 7 symbols out means a time delay of $7 \times 0.144 \times 10^{-6} = 1.01 \times 10^{-6}$ seconds. For a coax propagation coefficient of 0.8, the distance to an echo is the speed of light x coefficient x time = $3 \times 10^{8} \times 0.8 \times 1.01 \times 10^{-6} = 242.4$ meters. The bad news is that the reading does not tell you where the fault is exactly. For example, the fault could be 242 meters on the trunk, 70 meters off a branch 172 meters away, or a break anywhere in between.

Digital video depends on compression

Compression puts more in less

Without the aggressive developments in digital compression techniques, digital video would not be an industry. Not only does compression allow digital bit streams to fit in available bandwidth, but also the bandwidth savings is so great that the remainder can be used for error correction. At its most basic definition, **compression** reduces the volume of data by removing redundant adjacent pixel information in a single TV frame and redundant information from frame to frame.

Compression

Video compression is a digital process that enables transmission of digital video, reduces transmission bandwidth to make error correction possible, and permits the choice between high definition television (HDTV) or a number of standard quality channels in one channel.

Digital compression quality is tough to measure

When a subscriber complains of poor picture quality, you want to be able to troubleshoot quickly. Compression problems are usually beyond the control of your system because most signal compression occurs at the source of the digital stream, such as from a digital center delivered by satellite or long-distance optical fiber. Remodulation of a signal at your head end to change its format, multiplexing digital streams together into a single digital signal, or locally originated programming do not involve digital compression. Artifacts on a picture can be caused by poor compression at their source. But system impairments can also cause data bit stream problems which may be mistaken for compression artifacts, preventing required troubleshooting. For this reason it is wise to know something about compression in general and a little more about the most prevalent form of compression in the cable industry, MPEG-2.

Compression for cable television currently centers on the delivery of entertainment, but data, high definition television (HDTV), and other, not yet realized uses, are sure make compression as much a part of system design and maintenance as optical fiber technology is today. The following sections are an introduction to show the effects of compression on your day-to-day measurements.

Types of compression

Compression always loses something

Compression requirements vary between applications, storage media, transport methods, and video programming types. The factors that dictate the type of compression include timing, program data stream reconstruction, synchronization, demultiplexing/remultiplexing, packeting/repacketizing, and encryption. These topics are beyond the scope of this book, but it is well to know the main classifications of compression: lossy and lossless. **Lossy compression** is based on the concept that the information lost will not be missed because the human or machine using the reconstructed information is not capable or interested in seeing the loss. This can be a shortcoming when real-time transmissions change bandwidths too fast. An example is the transmission of a basketball game where the foreground and the background are moving relative to one another while the picture is panning to follow the action. Compression ratios of 200:1 to 400:1 are common, allowing a trade-off between signal quality and storage or transmission speed. Variable-rate compression is used on stored media to optimize the signal as it changes, while putting the most material in the least amount of storage.

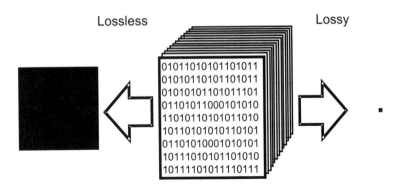

Figure 7.9. Lossless and lossy compression ratios represented by areas.

Lossless compression does not lose a bit of data, but it does compress by a factor of 2:1 to 4:1. Lossless compression is necessary when every last data bit is required by the receiver to operate properly. Compression is achieved by noting how often a specific sequence is repeated, not repeating the sequence for each instance. Examples of lossless compression are the transmission and storage of software applications and data files, such as games and financial spreadsheet information files.

Lossy and lossless compression	Lossless compression requires the received signal to be a duplicate of the original. Lossless compression ratios of 2:1 to 4:1 are common. Lossy compression allows the trade-off of bandwidth for quality. Lossy compression ratios of 200:1 to 400:1 are common.

The contrast between the lossy and lossless payload reductions is seen in Figure 7.9. The data payload is represented by the area of the square in the center. The lossless compression reduces to the area on the left, and the lossy reduces to the area on the right. It is easy to see why lossy compression makes transmitting digital video possible!

For video studio use, a near-lossless digital compression of baseband PAL, NTSC, and SECAM time-multiplexes three video components in a ratio that causes a slight loss in the color-luminance combination. The lost information is selected to coincide with effects difficult for the human eye to see. However, data rates still reach 200 Mbps.

Clearly this type of near-lossless format is only valuable for the short distances of a broadcast studio.

Compression in everyday life

We live in a world that uses many forms of compression to get information moving more efficiently. When two words are pushed together, with an apostrophe to represent the missing letters and spaces, called a contraction, it's in an effort to save space on the printed page or time in speaking.

Voice transmission in telephone, when modulated onto a terrestrial, cellular radio, or satellite carrier, is compressed using algorithms which take advantage of the limited number of sounds a human can comprehend, the nature of language phonics, and the limited bandwidth necessary to make sense of a voice.

Until the decline in computer storage and memory costs, many schemes for compressing software and data helped keep computer operation cost-effective. Today, compression is still popular to make large data files, such as graphics, compact enough to transmit economically.

Compression is part of analog television. Most of one half of the lower sideband is discarded at the transmitter, which you know as the vestigial sideband technique, to save transmission bandwidth. In addition, only half of every frame is updated at the refresh rate with the interlacing of horizontal lines. Bandwidth is saved because only half the picture information is sent for each frame, and the sedentary frame relies on the persistence of the television phosphors and memory of the human eye and brain.

Compression forms and formats

Compression for video is not new. The first practical television broadcast systems used a form of bandwidth reduction, called interlace. Instead of sending 50 or 60 frames per second, each frame is divided into two fields containing half the total number of lines. This rate is just fast enough to look like smooth motion, although some people claim to be able to see flicker in the 50-Hz PAL video.

There are compression standards for every use

Another form of video compression is used in color television as it translates the three color signals, red, green, and blue, into the color difference domain, where the picture is represented by luminance and two color difference pictures. The limitations of the human visual system, that is, less sensitivity to color than to changes in brightness, allows the bandwidth of the color signals to be reduced by half.

For digital pictures, the lossy JPEG format, primarily for still picture compression, requires too much bandwidth for use in motion transmission. Its technology is used in creating the lossy, but very high quality reference frames in the MPEG processing discussed below. Figure 7.10 shows two pictures compressed with the JPEG format. The ratio between their storage requirements are about 8:1, even though the quality of the expanded pictures is comparable. The difference in compression is due to the repetitive nature of the texture in the picture on the left in contrast to the very busy texture of the picture on the right.

Stored size: 184,000 bits   Stored size: 816,000 bits

Figure 7.10. JPEG-compressed still pictures having 640 x 480 pixels and a range of 256 colors.

There are compression formats for every video purpose, as Table 7.1 demonstrates. This information is from Minoli, listed in the bibliography at the end of this chapter. These are but a few of the more than dozen formats in use for various forms of video transmission and storage.

Compression Standard	Use	Approximate range of data rate	Compression compared to broadcast quality
ITU-R 601-1	studio-quality digital television	270 Mbps	reference
DVI/Indeo	multimedia and CD-ROM-based PC applications	1.2 to 1.5 Mbps	160 times
H.261	video conferences and video telephony over ISDN	64 kbps to 2 Mbps	24 times
Motion JPEG	such as medical applications	10 to 20 Mbps	7 to 27 times, with less quality of pixels and color
MPEG-2	such as digitized motion pictures, satellite transmission	3 to 10 Mbps	30 to 100 times
MPEG-2	high-definition television	15 to 20 Mbps	30 to 100 times

Table 7.1. A comparison of lossy video compression parameters to those of the studio-quality standard, ITU-R.

MPEG-2 compression uses

MPEG-2 is the current standard for the distribution of moving pictures used by most cable, satellite, and newer laser disk technologies. It is the product of an international group of scientists, engineers, and industrial/commercial enterprises whose central organization, ISO/IEC's Moving Pictures Expert Group (MPEG), provides and maintains the standards. The MPEG organization is in motion in another sense. They are carrying the technology forward to MPEG-3, 4, etc, to meet future needs of digital multimedia. The current users of MPEG-2 include the European Digital Video Broadcast standard (DVB), direct satellite broadcasters, and the Digital Versatile Disk technology (DVD), to name a few. With its list of successful users, and the continuing research and development of digital media technology, current and future MPEG standards are likely to meet the cable industry needs for the foreseeable future.

MPEG-2 advantages	**Adopting MPEG-2 as the standard for compression of video allows the designers and manufacturers of equipment to keep the costs to the end-user low. The standard also requires that much of the coding be done on the transmitter side, so that the receiver hardware and software costs can be kept low. This means lower-cost set-top boxes for your subscribers and fewer troubleshooting problems for you along the transmission route.**

MPEG-2 is independent of modulation format

The symbol rate of a signal translates roughly into transmission bandwidth, not withstanding filtering effects. The faster the symbol rate, the wider the bandwidth necessary to transport it without degradation. To save precious bandwidth, while keeping up with the transmitted channel information, MPEG-2 compression occurs early in the transmission process, that is, as it comes out of the studio or other baseband source. Compression is done at digital baseband, after the source signal has been digitized, but before the signal is formatted into its transport data packets, error corrected, preemphasized, and modulated. The MPEG-2 process is independent of the type of error correction, preemphasis, and the digital modulation format chosen. MPEG-2 is used with VSB and COFDM in over-the-air broadcast, and several versions of QAM for cable distribution.

Compression processes, including MPEG-2, use specific mathematical rules called algorithms to reduce the volume of data by

- eliminating duplicate information in the same frame
- eliminating information that does not change from frame to frame
- reducing information that the human senses cannot observe or may not miss

The MPEG-2 algorithm is asymmetrical in addition to being lossy, that is, it requires more computational complexity (hardware and software) to compress full-motion video than to decompress it. This means that the high investment in operating equipment for distribution systems is in the source side. For cable television this means the head end sources of programming material, such as digital center from a satellite or fiber link, do all the compression work. Neither the head end nor the receiver of the compressed material has to be packed with processing hardware and software. The subscriber's set-top box can be lower cost, standardized, and simple.

How MPEG-2 works, in brief

The video compression for MPEG-2 is a complex process, but here are the highlights. First, the process does not transmit pixels in a frame that do not change. Second, frames with redundant information are compressed by eliminating duplication by lossless compression algorithms. Third, complete single frames, called intraframes, or I-frames, are transmitted only about once out of every 15 frames. These frames are compressed using the JPEG still-picture compression format. Fourth, the motion between I-frames is created from the information supplied by the I-frames. The frames that capture motion are the P-frames and B-frames. P-frames are the prediction frames which reconstruct a new frame from the previous P-frame or I-frame plus the data transmitted just after the last frame's completion. Objects in motion continue along the same path, with backgrounds patched in from previous frame data. The encoding process precedes the transmission of several frames, making the B-frame possible. The B-frame, where B is for bidirectional, also compresses motion. It is encoded similarly to the P-frame, except that it can choose patching information from either the last frame or the next constructed frame, depending on which "looks" better to its built-in quality algorithms.

If the picture is not changing, only compressed frames are sent using a JPEG format, and the motion frames are not used. If the only thing on the screen moving is a ball bouncing across the floor, the motion compression comes into action. But only the object in motion, and the background being exposed are being transmitted. The unchanged pixels are only sent occasionally. If everything on the picture is moving, the compression begins to work a bit harder. And if all pixels are changing, as when many objects are moving relative to one another, as in a basketball game played on a patterned floor, then the compression data stream needs its widest bandwidth.

For an in-depth explanation of compression in general and MPEG-2 specifically, see the references Orzessek and Sommer, Minoli, and *Interval*, the SCTE Newsletter in "Selected bibliography" on page 145.

Trouble from compression

As mentioned, encoding usually takes place long before the signal reaches your head end. It is buried deep in the signal's data structure so that picture problems, called artifacts, caused by compression encoder failure, are not often affected by the BER and other data stream quality influences. Interruption in the data stream usually causes blocking and blanking in the picture. This is illustrated in the top two frames of Figure 7.11. Compression encoder problems are more likely to look like the bottom two pictures of the figure, with blurring of the objects, and motion artifacts such as trails and color changes.

Compression failure is probably not within your power to fix, unless it is the decoder at the set-top box that is at fault. But the number of trouble calls should tell you if the problem is spread over a wide area, confined to a particular hub or branch, or in one subscriber's drop.

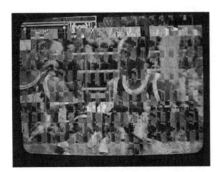

Figure 7.11. Compression breakdown effects on TV pictures.

Compression measurements

The process of compression encoding does not cause a modulated signal to crash. The picture quality suffers the effects mentioned in the last section, but the compression process simply loads the data stream with bits created by the compression coding, just as any other signal, compressed or not, gets transported.

Variable-rate compression can cause crashes

MPEG encoding, and therefore the data stream, is dynamic; it changes compression and data stream bandwidth to fit the programming. This variable-rate compression allows dynamic usage of bandwidth based on data content. For the compression of prerecorded material, the advantage is that the compression can be tailored to the material, to give the playback the best quality possible for the compression standard and available bandwidth. For the distribution of real-time, live-action material, such as sports, the distribution system must be able to respond to the changing bandwidth needs as required. If an MPEG-2 stream is multiplexed with other bit streams, MPEG rate changes may cause the multiplexed stream to go over the cliff.

Some multiplexed signals are coded with program identifier tags, called PIDs, that can be monitored by protocol analyzers, from the demodulated data stream, and digital modulation signal analyzers prior to demodulation. PID bandwidth is a statistical measure to detect anomalies in the transmission, to show encoding bit rate problems.

Program identifiers watch data stream loading

When data streams of different baseband digital data streams are multiplexed together, the information is tagged with identification. The streams are multiplexed together in alternating bursts of data called packets. Each packet is assigned an identifier to show which program it belongs to. Measurement equipment such as protocol analyzers can monitor the content of the data stream by reading these identifiers. For compressed signals from MPEG, these identifiers are called program identifiers, or PIDs. PID values are analogous to the identifiers in other data stream transport systems such as frame relay, ATM, and X.25. In the other transport systems the identifiers are known by other, equally cryptic names or acronyms. The use is just the same: to identify which packets belong to a specific packet stream.

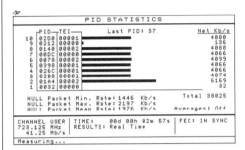

In the Hewlett-Packard QAM Analyzer, program identifier (PID) statistics provide two critical pieces of information about the MPEG transport stream. First, they list the individual data streams which make up the transport stream, along with the data rate associated with each one. An operator can check the video, audio, and data content of the stream, and check for "illegal" PIDs. The number of "null" packets (added to bring the overall data rate up to the rate used for the channel) determines how much more information could be added to the stream. Second, they allow identification of any PID which has suffered uncorrectable packet errors during transmission. Packet errors cause glitches or blocking effects on the picture.

Summary

The layers of signal processing packed into a digital signal include compression, error correction, and adaptive equalization. Each transmitter has a reciprocal receiver process to restore the signal to its original form. DBV-C in Europe is a digital video standard which specifies the exact formats for these and other processes.

The causes of errors in digital signal transmissions are similar to those that cause degradation to analog signals: frequency response, delay distortion, signal-to-noise ratio, harmonic distortion, frequency instability of the signal or clock (jitter), impulse noise interference, and echo (reflections).

Bit error rate, BER, is the ultimate quantitative measurement parameter for digital signal quality, but is being replaced by terms more suitable to the type of data transmitted. Frame and packet error rates are common, but BER is useful for instructional purposes. BER is the ratio of transmitted bad bits to good ones.

Error correction tries to keep the received digital signal at its peak quality against all adversity. Forward error correction, FEC, sends redundant information along with the signal to allow it to be repaired at the receiver if necessary. FEC takes signal bandwidth away from the main payload.

FEC masks a degraded BER at the output of the receiver, making it difficult to judge how close to the cliff edge the signal is operating. Margin is the term given to the guardband between the current operating point and the cliff edge. The availability of FEC statistics, such as Reed-Solomon coding, can show how hard the error correction mechanisms are working. Statistics on real-time signals require dedicated equipment and standard measurement guidelines, such as provided by the European DVB-C standards for digital video signals in cable distribution.

Adaptive equalization compensates for transmission impairments by removing linear distortion effects from the signal. Linear distortion is caused by tilt, ripple, and group delay. Group delay causes distortion when signals of different frequencies arrive at the receiver with delays long enough to distort the signal's waveform. Digital signals are specifically susceptible to group delay distortion, which is caused by reflections and phase shifts from physical and structural distribution problems. As with FEC, genuine network problems can be hidden from the test engineer when equalization is used.

An adaptive equalizer is a digital filter that continuously reprograms phase and amplitude characteristics to compensate for the phase and amplitude distortions of the transmission channel. As such, a look at how bad your system is behaving can be seen by monitoring the receiver's equalization frequency and phase responses.

Compression reduces digital information by 2:1 to over 400:1. Without compression, the common distribution of digital video would not be possible due to the costs of transmitting the wide bandwidths of uncompressed signals. Because compression and decompression are handled mostly outside the typical cable television system, there are few procedures available to correct compression problems. It is sufficient to identify them in the cable system.

MPEG-2 is the current standard for distribution of moving pictures and is likely to endure because of current success and aggressive design and research. successors.

Compression is a coding process independent of FEC, equalization, and modulation. Like other compression techniques, MPEG reduces the transmission bandwidth by eliminating duplicate information in a single frame and information that does not change from frame to frame. It also removes picture information that humans will not miss.

The picture and sound problems when compression fails is different from a general failure of the bit stream. Bits stream failure is often blackout and blocking. Compression failure is usually motion artifacts such as trails, blurring, and stop-and-start movement.

Questions for review

1. For each of the following, tell whether the process is carried out on the modulated signal, the digital baseband signal, or both.
   a) forward error correction
   b) adaptive equalization
   c) compression

2. What is the purpose of error correction? Choose all that apply.
   a) to correct transmission bandwidth for sharp roll-off
   b) to supply bits damaged in transmission
   c) correct spelling in your compliance reports
   d) to keep the signal quality as high as it can be
   e) to keep you from seeing how close to the cliff the signal is

3. What is the purpose of equalization? Choose all that apply.
   a) protect the signal against microreflections in the system
   b) to protect the signal from linear distortions in the system
   c) to protect the signal from non-linear distortions in the system
   d) to regulate taxes on your cable system

4. What is the purpose of digital compression? Choose all that apply.
   a) to save transmission bandwidth
   b) to reduce the amount of data sent while maintaining the highest signal quality possible
   c) to make it possible to transmit high-quality video signals

5. The cliff effect is the tendency of a digital signal with error correction and adaptive equalization to maintain high quality even as impairments increase until it
   a) ceases to be transmitted
   b) cannot be processed by the receiver
   c) slowly degrades the signal

Selected bibliography

1. A Guide to Picture Quality Measurements for Modern Television Systems, Tektronix, Inc., Internet address: www.tek.com/measurement/App_Note/PicQuality/picture.html, February 1998.

2. "Field Installation and Maintenance Testing of DVB Systems," a paper for Hewlett-Packard Company System Engineering Training, Microwave Instruments Division, October 1996.

3. *Interval*, the SCTE Newsletter, Society of Cable Telecommunications Engineers, Inc. *DigiPoints*, June 1997

4. Helen Chen, "Testing Digital Video: A Look at Measuring Power and Interference," Communications Technology magazine, Phillips Business Information Inc., Rockville MD, May 1995.

5. John Ernandez, "Understanding the Effects of Noise on Digital Signals", Communications Engineering & Design, Chilton Publications, New York, December 1997.

6. Brian Evans, *Understanding Digital TV; the Route to HDTV*, ISBN 0-7803-1082-9, IEEE Press, Piscataway, New Jersy, 1995.

7. Dr. Kamilo Feher and the Engineers of Hewlett-Packard Ltd., *Telecommunications Measurements, Analysis, and Instrumentation*, ISBN 0-13-902404-2 025, Prentice-Hall, Inc., Englewood Cliffs, New Jersey, 1987.

8. Daniel Minoli, *Video Dialtone Technology, Digital Video over ADSL, HFC, FTTC, & ATM*, ISBN 0-07-04-2724-0,McGraw-Hill, Inc., New York, 1995.

9. Harry Newton, *Newton's Telecom Dictionary*, 12th Edition, IBSN 1-57820-008-3, Flatiron Publishing, Inc., New York, February 1997.

10. Nicholas Negroponte, *Being Digital*, ISBN 0-679-43919-6, Random House, Inc., New York, 1993.

11. Michael Orzessek, Peter Sommer, *ATM & MPEG-2, Integrating Digital Video into Broadband Networks*, Hewlett-Packard Professional Books, ISBN 0-13-243700-7, Prentice-Hall PTR, Inc., Simon & Schuster Company, Upper Saddle River, New Jersey, 1998.

12. Dragos Ruiu, "The Challenges of Compressed Digital Video," 1996 Digital Video Test Symposium Technical Paper, Hewlett-Packard Company, 1996.

13. Dragos Ruiu, "An Overview of MPEG-2," 1996 Digital Video Test Symposium Technical Paper, Hewlett-Packard Company, 1996.

14. Wayne Smith, paper for the HP 89410A/41A Vector Signal Analyzer field training event, Hewlett-Packard Company, June 1996.

15. John Watkinson, *The Art of Digital Video*, ISBN 0-240-52369 X, 2nd Edition, Focal Press, Oxford, 1995.

16. Robert A. Witte, *Spectrum & Network Measurements*, ISBN 0-13-030800-5, Prentice-Hall PTR, Inc., Simon & Schuster Company, Upper Saddle River, New Jersey, 1993.

8

Digital Signal Quality

Signal quality is influenced by many system and signal characteristics. This chapter discusses those influences and defines the parameters and measurement techniques used to define signal quality.

What you will learn

- What affects digital signal quality?
- How is digital signal quality measured?
- Bit error rate measurements
- What role does digital modulation play in signal quality?
- Troubleshooting tips using constellation and eye diagrams
- What are EVM, MER, and SER?
- Can margin be measured?
- What role do jitter and clock timing have in signal quality?

Why measure signal quality?

Try to operate with plenty of margin

Subscriber complaints are not enough to tell you whether or not your quality is high enough. When a digital signal fails, it is usually with a crash, not a whimper, so your customer is likely to be very unhappy, just as they are when a construction crew cuts through a trunk during the "big game." With digital signals it is important to monitor actual signal quality, not delivered signal quality as assurance that the signal and system are operating with plenty of margin.

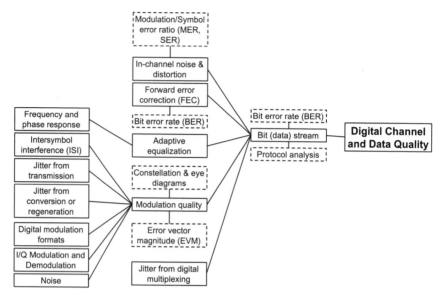

Figure 8.1. Influences on digital channel and data quality.

What degrades digital signal quality?

Just about every aspect of the transmission path affects signal quality. Maintaining your system for good analog signal quality is an excellent starting point. But digital signal quality depends upon some additional factors, as shown in Figure 8.1. The quality goal, on the far right, means few, or no dropped bits to subscriber application software and consistent picture quality for their video entertainment. Only one measure, the bit stream, feeds into digital signal quality. BER is the best, and only, true measure of end-to-end digital transmission quality. The columns to the left show that the bit stream quality is dependent upon quite a few cascaded factors. The solid boxes are the system performance attributes, and the dashed boxes are the measurement parameters or techniques. In Figure 8.1, the attributes in the solid boxes can be measured if they have an attached parameter box, noted with the dashed lines. You can use these parameters to monitor performance and troubleshoot system problems.

Figure 8.2 is another way to view signal quality factors. It shows the general location of the leading causes of signal degradation in a simplified system diagram.

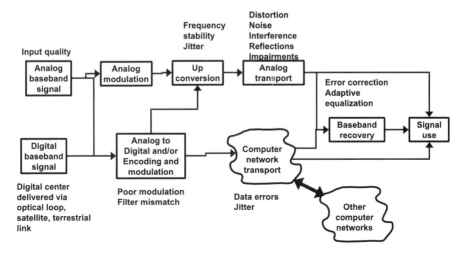

Figure 8.2. Factors of system signal quality.

These characteristics are measured with techniques that can be grouped into three categories, the first two covered in this chapter:

1) Data quality, end-to-end through the cable system.
2) Modulation quality parameters, many of which can be used to troubleshoot sources of power, distortion, noise, and interference.
3) Power, distortion, noise, and interference measured along the transmission path. See Chapter 9 beginning on page 155, Chapter 10 beginning on page 191, and Chapter 11 beginning on page 211.

Qualitative = troubleshoot Quantitative = specs

When discussing measurements, it is important to know when a measurement is considered qualitative or quantitative. **Qualitative** testing lets you observe the digital signal in progress, watching for unusual patterns that may be indications of specific impairments. With a qualitative test, there is no number you can use to compare to a specification or standard, but the results often point out trends or hints for troubleshooting. Qualitative testing, also called subjective testing, provides information that puts your perception and experience to work to check signal quality. Constellation and eye diagrams are qualitative tools.

DVB-C digital video broadcast standards for cable

DVB-C in Europe is a good example of standardized digital system quantitative specifications, procedures, and hardware. An example of a DVB standardized specification is modulation error ratio, or MER. The DVB Measurements Group document, ETR290, states that MER shall be used as the "single figure of merit" for signal quality in a DVB system. As digital signals become common in cable systems, compliance to standards such as for the DVB will become important to sustained growth and economy of scale, just as the American FCC proof-of-performance has in North American cable systems.

Quantitative measurements provide a number that can be compared to a specification. Bit error rate, covered next, is a quantitative measurement. If you are in North America, you make quantitative analog cable performance measurements daily. Specifications such as C/N, CBT, CSO, and carrier power are quantitative measures. For digital signals, specifications such as MER, EVM, and margin are being standardized.

Bit error rate

Bit error rate, or **BER**, is a quantitative overall measure of the data quality of the received bit stream. It is the ratio of the number of bit errors to the total number of bits sent in a given time interval. A bit error occurs when a "1" or "0" is interpreted by the receiver as the opposite value. As the error rate increases, the quality of the signal's data is reduced until he receiver can no longer make sense of the message. For video, this causes the picture to fail. For data, even small error rates cause application failures because every bit is needed by the receiver to reconstruct software code, such as spreadsheet, game, or executable software files. Small errors generated in video and audio can often be overlooked because of their fleeting nature. For compressed data, such as MPEG, small data losses are noticeable because the data and the instructions for their decompression may be missing.

Data are more sensitive to bit errors than video

Bit error rate

Bit error rate, or BER, is the ratio of the number of bit errors to the total number of bits sent in a given time interval. Bit error rate testing is one sure way to see the signal quality through your system. But BER end-to-end masks the corrections automatically made by adaptive equalization and error correction, so it is not a good indicator of how stressed your system is, that is, how close its operation is to the cliff edge.

Error correction can hide the true signal BER. Since signal impairments do not affect picture quality unless they cause unrecoverable bit errors, the only place to see the "raw" signal is at a test point between adaptive equalization and error correction decoding layers in the receiver. If your system includes set-top boxes that allow access to this point, the measure can be helpful in determining margin.

More commonly, BER is used as a quantitative measure of end-to-end system performance. If BER is measured from end-to-end, that is, from the source of the baseband signal to the output of the receiver, it includes the error correction processing and provides a complete check of transmission and decoding quality. Often BER is measured several layers of signal processing away from the direct picture quality, but is still a reliable measure of quality. In end-to-end testing, the BER value reflects only impairments severe enough to cause bit errors, remaining insensitive to subtle trends in the digital modulation. A good BER indicates proper service delivery. A bad BER highlights impaired service, but does not identify the cause of the problem.

"Good" BER

The definition of good BER depends entirely on the digital format of the transported signal. Every digital system has a minimum BER required to deliver the specified signal quality to the receiver. These levels are dependent upon the type of modulation, the definition signal quality, and a number of other parameters. For example, a BER of 1×10^{-6} is considered a threshold for visible degradation of a 256-QAM signal without error correction. Forward error correction can improve BER from 10^{-4} before FEC to 10^{-12} after. In DVB-C digital video, a BER of 10^{-10} produces no visible degradation of picture quality.

Chances are that your subscribers will see the effects of poor BER before you. Here are some of the indicators of poor BER they may report.

BER does not reveal the source of errors

- low speeds in data links, including cable modems, due to retransmission of garbled data
- digital video picture tiling from dropped compression bits
- digital audio static and noise from dropped data and compression bits

Unfortunately, BER itself does not point to the source of the error, so troubleshooting must be done using other measurement tools, such as constellation diagrams, or techniques, such as component substitution.

Measuring bit error rate

There are two types of BER measurement techniques: intrusive and nonintrusive. Intrusive BER injects a signal at the front of the down-converter sampler, and a bit error rate tester (BERT) is placed at the output of the FEC. Standard pseudorandom bit sequences (PRBS), standardized on a world-wide basis, are used so that the BERT knows what to look for. The test is intrusive because the video baseband is replaced by the test bit sequence. Unlike analog test signals, digital data cannot be inserted into digital video as easily as they are placed in the blank NTSC and PAL video lines.

BER measurements require test planning	Bit error rate testing requires careful consideration when designing a system's digital architecture. You need to decide early in the design whether the data paths can be tested sufficiently with either intrusive or nonintrusive techniques. The type of system equipment as well as the access points need to be built into the system design from the beginning; retrofitting BER capability is awkward and expensive.

As mentioned, bit error rate is typically an end-to-end system measurement. To use a BERT, which operates on the demodulated digital baseband, a digital test receiver is required. The measurement setup is shown in Figure 8.3. The digital test receiver monitors both sides of the transmission channel, shown as points A and B, and acts as a digital video receiver to recover baseband digital streams for the BERT. In addition to the demodulated bit stream, the BERT may need a bit clock input, although often it can recover the clock information from the data stream.

To perform this test, you need to decide what layers of processing to include in the test receiver. Do you want to compare BER before and after RF transmission? If so, both A and B test points must be designed into your system. It is important to consider testing strategy when designing the digital signal path. In the simplest version of a BER test, the known bit stream passes through the transmitter, bypassing the FEC. The BERT measures the received bit stream after being demodulated and decoded in the digital cable receiver. Since the known bit stream displaces cable TV program data, this is an intrusive test.

BER testers have three common PRBS bit-pattern lengths: 63, 511, and 2047. The patterns, which can be repetitive and simple, or random and complex, are sent through the system under test to allow a statistical reading of the BER. If the BERT is designed for a specific digital video receiver, such as a set-top box, your only indication of pass or fail may be a light indicator, not a BER value.

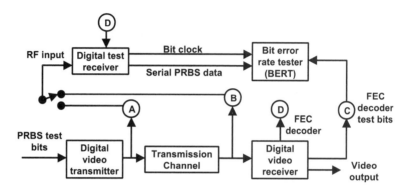

Figure 8.3. Setup for measuring BER.

Nonintrusive BER testing is done using the error information in the receiver's forward error correction chip, test point D in Figure 8.3. Since the FEC decoder recognizes bit errors and corrects them, the decoder counts and records the number of bit errors that have been corrected. The FEC chip can supply an accurate BER without injecting a known bit stream. This nonintrusive BER depends on the power of the FEC to detect bit errors. Different FEC codes have different error detection sensitivities. A Reed-Solomon code, for example, can correct bursts of bit errors, as well as single bit errors. If the BER exceeds the error correction power of the FEC, the error statistics do not accurately reflect the true BER, so applying this method requires understanding the FEC limits.

BER: catch at least 100 errors for a good measurement

Regardless of whether the BER testing is intrusive or nonintrusive, or with or without FEC, it is important to gather enough information to make a BER test statistically significant. A rule-of-thumb for BER measurement accuracy is to continue testing until at least 100 errors have been detected.

Accurate BER testing may take a long time

Accurate BER testing may take a long time. Since BER is a ratio of the bad bits to the total number of bits, a BER of 1 means that all the bits in a given bit stream are bad. A BER of 0.1, or 10^{-1}, means the every tenth bit is bad, that is, one out of ten. Forward error correction can produce BERs as low as 10^{-12}, which means only one bit out of 10^{12}, or 1,000,000,000,000 bits is bad. BER is measured over time since the bits flow in a stream at a given rate. At a bit rate of 6 Mbps, how long would it take to see a single bit error for a BER of 10^{-12}? The test time is the number of bits divided by the rate, or $10^{12}/6 \times 10^6$ = 1.7×10^5 seconds, or about 46 hours! If the BER is truly 10^{-12}, then the total time to catch 100 bad bits required for accuracy would be 4600 hours. More commonly, however, statistically accurate BER measurements on lower BERs only take several seconds to several minutes.

Use a BERT for replacement troubleshooting

A BERT is most effective in troubleshooting when a replacement device under test is available to substitute for the one in question. The BERT will quickly determine if the substitute has fixed the problem.

Viewing modulation with constellation and eye diagrams

Constellation and eye diagrams are qualitative digital modulation measurements that help you see what trouble is causing high BER or subscriber complaints. Constellation diagrams show the digital modulation symbol patterns, and eye diagrams show the symbol transitions. Details about how they view digital formats, and their relationship to one another are discussed in "Constellation and eye diagrams" on page 88.

Constellations and eyes	Constellation and eye diagrams measure the quality of the baseband digital modulation, and often let you see the source of the degradation.

To make a constellation and eye measurement on a head end digital video signal, you need a digital test receiver dedicated to the system's specific modulation format. See Figure 8.4. With this setup, the signal quality at the head distribution point is compared to subscriber reception. A test receiver with a digital modulation analyzer or a high-speed oscilloscope can be used. Measurements done at the output of the head end, point A, give a quick check on the quality of the digital video transmitter before cable channel distortions. If the test receiver can apply adaptive equalization, the linear distortion of the channel can be removed, permitting clearer views of baseband modulation impairments caused by noise, compression, and spurs at the set-top receiver, point B. For modulation formats with a great number of points, such as 64-QAM and 256-QAM, very high display resolution is required by the oscilloscope or analyzer to see the patterns.

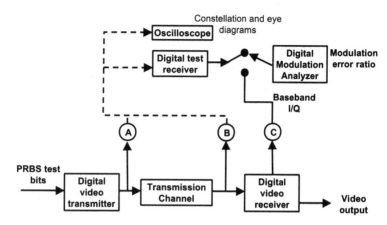

Figure 8.4. Test setup for viewing digital modulation and measuring modulation error ratio.

Troubleshooting with constellation diagrams

Constellation plots are graphical views of baseband digital modulation. Troubleshooting with a constellation diagram is a matter of recognizing a pattern that deviates from the ideal and applying a little common sense. If each measured constellation point corresponds with the correct ideal target, the signal is perfect, and there are no bit errors in the transmission. Non-ideal design, ingress, component aging, and many other effects conspire to degrade the signal and can be observed on a constellation diagram. Here are several common diagnostic illustrations from a Hewlett-Packard DVB-C QAM Analyzer, model HP8594Q.

Figure 8.5. Constellation diagram of a "clean" 64-QAM signal.

Clean signal: As a reference, Figure 8.5 shows the constellation of a clean DVB-C 64-QAM signal. The horizontal scale is I and the vertical scale is Q. The analyzer is displaying 800 symbols at a 6.9-MHz rate. The channel bandwidth is 8 MHz and the

signal is at 306 MHz.

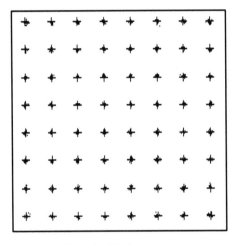

Figure 8.6. Amplifier gain compression on a constellation diagram.

Gain compression: When an amplifier goes into compression it loses its ability to amplify higher amplitude signals linearly, that is, the same output changes for every input increment. The constellation of Figure 8.6 shows the effect as a "pulling in" of the points having the highest power, the ones most distant from the center, or zero amplitude, point. The perimeter of the constellation becomes bowed as the corners are pulled in.

Compression is often caused by excessive signal noise loading, which may be the summation of high system input levels, ingress, and distortion signals at the offending amplifier.

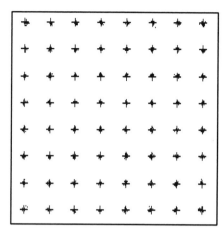

Figure 8.7. Gain imbalance of the system's components.

Gain imbalance: IQ imbalance is caused by different amplitude and phase distortions in the I and Q paths. In Figure 8.7 you see the horizontal symbol pointed squeezed in caused by 2 dB of gain imbalance. The vertical points are not as affected. Other measurement parameters, such as error vector magnitude, discussed later in this chapter, provide a quantitative measure of this signal degradation.

Imbalance indicates problems in the I/Q modulators, I/Q demodulators, or baseband amplifiers and filters. If I and Q are affected equally, look for trouble in IF amplifiers and filters, RF amplifiers and filters, up/down converters, clock recovery circuits, IF equalizers, and components of the communication systems.

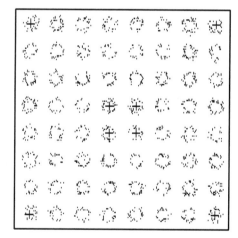

Figure 8.8. Spurious tone in the signal.

Spurious CW tone interference: An interfering sine wave tone, sometimes called a spurious interference, or spur, it is often not visible in the spectral display because its energy is hidden by the modulated signal's power. Figure 8.8 reveals a 200 kHz spur as it modulates each of the symbol points with a circular motion. The interference tone is 23 dB lower than the digital signal's average power. The size of the ring is directly related to the level of the interfering tone. Unfortunately, this display does not reveal the spur's frequency. To see the spur using a spectrum analyzer, use a narrow video filter to expose carrier leakage. If you can turn the channel off, the spur can be measured directly as a CW signal in the empty channel band. This is rarely practical, except at installation.

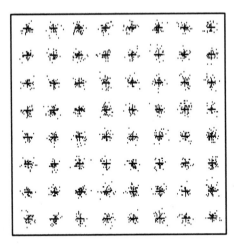

Figure 8.9. Noisy signal.

Signal noise: White or Gaussian noise caused by poor carrier-to-noise in the system causes the constellation points to disperse around their ideal locations as the noise adds random amplitude and phase to the each symbol trajectory. Figure 8.9 shows noise that is 18 dB below the total signal power. Such noise can also be the result of noise-like third-order distortion. See Distortion, Noise, and Interference, Chapter 11 beginning on page 211.

Figure 8.10. Carrier leakage.

Carrier leakage: Carrier leakage, also known as local oscillator feed through, or constellation offset, indicates an imbalance in the modulator's mixer or an unwanted DC condition in the transmission system. The effect of carrier leakage is an offset in the whole constellation from the expected origin. The magnitude and phase of the offset are directly related to the amplitude and phase of the interference. Figure 8.10 shows carrier leakage 30 dB down at an angle of zero degrees from the carrier. The offset may not be picked up by quantitative measurements, such as error vector magnitude, if the analyzer measuring the signal has been programmed to offset the leakage temporarily.

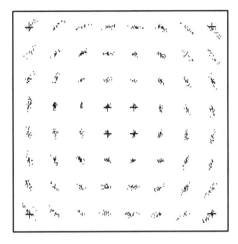

Figure 8.11. Residual FM.

Residual FM: The short-term instability of the signal's carrier can do great harm to digital transmissions by causing amplitude distortion to the waveform as the frequency shifts randomly. Frequency shifts cause the signal to shift in phase relative to the carrier, giving the constellation diagram in Figure 8.11 a distinct circular look. The most likely source of residual FM is from the instability of the local oscillators of upconverters and downconverters along signal's path. Converters are found at the head end for moving channels around in frequency and for satellite signal downconversion in receiver low-noise broadband amplifiers.

Always use qualitative and quantitative tests together when possible

In each of these cases a look at both quantitative measures and the qualitative pictures help you fully evaluate the signal's digital modulation quality.

Eye diagrams allow baseband modulating signals (I,Q) to be observed independently. Even though they provide a quick look at the convergence of the signals at sequential symbol points, eye diagrams are not usually included in dedicated digital video analyzers. They are more common in digital microwave link analyzers and digital modulation analyzers used for research and development. See Appendix C, Equipment for Testing Digital Signals in Cable TV Systems.

Eye closure

Eye closure is another way to "look" at the subjective quality of the modulation baseband. The wider the eye, the tighter the crossover points, as illustrated by the two diagrams at the left. In the top figures, the constellations diagram has considerable noise at each symbol point. In its companion eye diagram, these widened points fill the vertical spaces at the symbol transitions, closing the eye. The lower diagram shows much less closure because the symbol transitions are more tightly clustered.

Remember that the eye diagram is for either the I or Q signal of the modulation baseband, so the closure of one may not reflect the closure of the other, as in the I/Q imbalance example of Figure 8.7 on page 137.

Modulation quality measurements

The measurement of modulation quality summarizes the effects of distortion, interference, and noise all in one quantitative number. **Error vector magnitude**, or **EVM**, and modulation error ratio, or MER, are the common quantitative measures of digital modulation quality.

Digital modulation quality parameters

Analog video signal quality is measured by such analog parameters as depth of modulation, chroma-to-lamina-delay-inequality, and carrier-to-noise ratio. The signal is analog, as are the measurement parameters. Constellation diagrams provide a look at the qualitative quality of the digital modulation baseband of a signal using analog signal processing. The constellation I/Q information can be further processed by the measuring equipment to provide many different qualitative measures. Here are the most common parameters:

- Error vector magnitude, EVM
- Modulation error ratio, MER, also known as signal-to- noise ratio, or SNR
- I-Q error magnitude
- Magnitude error
- Phase error
- Frequency error
- I-Q offset amplitude droop

Of these, MER and EVM are covered in detail because they embody the best of all the quantitative measurements. There are countless others, especially as you take into account the ever-expanding list of digital formats and systems for cable, optical fiber, and wireless communications.

- Error vector magnitude is a measure of digital modulation quality which uses a ratio between the amount of fuzzy spread of a symbol cluster and the average power of the symbol itself.
- Modulation error ratio is a quantitative measure of how well the symbols can be read against the random noise in the channel. It is similar in nature to signal-to-noise ratio, SNR, and often annotated as such.

Digital signal quality

Modulation of a digital signal onto a carrier is affected by many things. Looking at the quality of the modulation along the transmission path can help you find some problems, or eliminate suspected causes.

Error vector magnitude

EVM: How fuzzy is the symbol cloud, in %

Error vector magnitude measures modulation quality by assigning numbers on the fuzzy constellation clusters seen in the impairment examples in "Viewing modulation with constellation and eye diagrams" on page 134. Figure 8.12 shows this graphically. The constellation of the left, with its four symbols is quantified by comparing the amount of RMS error magnitude to the average magnitude vector. The EVM measure quantifies the deviation from the ideal symbol position no matter what type of impairment, shape, or position of the dot cloud. It gives a measure of the size of the clusters of symbols about the target symbol points on the constellation, thereby providing both a simple, quantitative figure-of-merit for a digitally modulated signal.

EVM is written into many digital communications standards of test, including wireless applications for GSM, NADC, and PHS, but is not part of the formal European DVB Measurements Group Standards. DVB has adopted modulation error ratio as their primary figure-of-merit for in-channel noise ratio measurements. See "Modulation error ratio" on page 144.

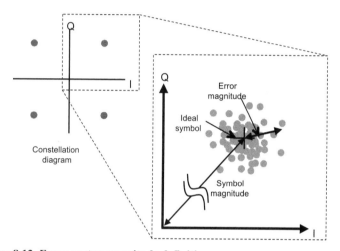

Figure 8.12. Error vector magnitude definition.

A simple definition of EVM is

$$\text{EVM} = \frac{\text{average error magnitude}}{\text{symbol magnitude}} \times 100\%$$

where EVM is given as a percentage of the error magnitude to the symbol magnitudes. Different standards use different techniques for quantizing the magnitudes. For example, error magnitude may be

Good EVM   **Levels of EVM of 2% and below are usually acceptable for QAM signals in DVB systems and in North America.**

measured as the RMS value, peak value, or average. The symbol magnitude may be the maximum or peak. Each one is clearly specified for the specific modulation and system standard.

Residual measurements

EVM is an example of a residual measurement, one which removes the signal under test and displays the differences between absolute parameters. This technique reveals small errors for further scrutiny. In the analog cable proof testing tests, the hum measurement is a residual measurement since it compares the amplitude of the carrier to the amplitude of the hum level. By providing a ratio between two parameters with the same units, the test value is unitless. Hum and EVM are multiplied by 100 so that percent can be tagged onto the numbers.

Measuring EVM

EVM is measured by modulation

The original measure of digital modulation quality (for QPSK and QAM modulated signals) was eye closure for the I and Q perspectives. (See "Eye closure" on page 139.) Eye closure has been replaced in most digital modulation industries by MER and EVM.

EVM measurements are made by test equipment that looks and acts like the system receiver in your subscriber's living room, the set-top box. For field work, specialized analyzers, such as the **QAM analyzer**, demodulate the RF digital signal, and apply the same FEC, adaptive equalization, and perhaps decompression as is done in the set-top box. To see the constellation, graphs and EVM results are measured at test point C in Figure 8.4. The analyzer compares the I and Q deviations from ideal symbol levels, and presents the ratio of peaks, averages, or RMS values as required. Figure 8.13 shows the constellation and EVM results from a Hewlett-Packard DVB-C QAM analyzer. The analyzer demodulates the transmitted signal prior to forward error correction, equalizes the signal, and calculates the average size of the error vector in relation to the maximum magnitude at a given symbol. The percentage result is shown in the bottom of the constellation diagram.

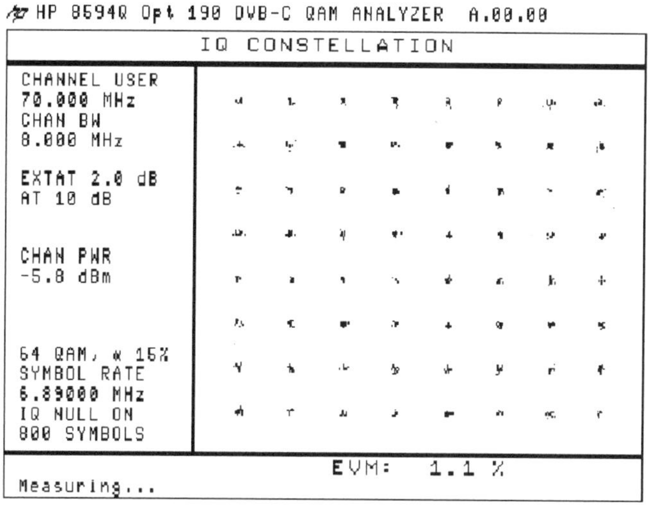

```
HP 8594Q Opt 190 DVB-C QAM ANALYZER  A.00.00
┌─────────────────────────────────────────────────────────┐
│                   IQ CONSTELLATION                        │
├────────────────┬──────────────────────────────────────────┤
│ CHANNEL USER   │                                          │
│ 70.000 MHz     │                                          │
│ CHAN BW        │                                          │
│ 8.000 MHz      │                                          │
│                │                                          │
│ EXTAT 2.0 dB   │                                          │
│ AT 10 dB       │                                          │
│                │                                          │
│                │                                          │
│ CHAN PWR       │                                          │
│ -5.8 dBm       │                                          │
│                │                                          │
│                │                                          │
│ 64 QAM, α 15%  │                                          │
│ SYMBOL RATE    │                                          │
│ 6.89000 MHz    │                                          │
│ IQ NULL ON     │                                          │
│ 800 SYMBOLS    │                                          │
├────────────────┴──────────────────────────────────────────┤
│                       EVM:   1.1 %                        │
│ Measuring...                                              │
└─────────────────────────────────────────────────────────┘
```

Figure 8.13. Measuring EVM with a DVB-C QAM digital modulation analyzer.

There is a growing choice in the test instrumentation offered for digital signal quality measurements, many combining a number of test-parameter capabilities. Besides the QAM analyzer, the **vector signal analyzer** is used to demodulate and analyze a variety of digital modulation parameters besides EVM. EVM vs. time, error spectrum, frequency response, eye and constellation diagrams, residual FM, and phase noise can all be seen, often on the same screen so that interactions can be observed. The vector signal analyzer is primarily a design tool for research and development, but has increasing value in the high-performance head ends and digital formatting centers.

A vector signal analyzer display is shown in Figure 8.14. EVM percent is listed in the table of text, but is accompanied by displays of EVM over time, the constellation and eye diagrams, and a number of other parameters. See "Digital video modulation analyzer" on page 255 and "Vector signal analyzer" on page 260 for more information on the measurement equipment and their test parameters.

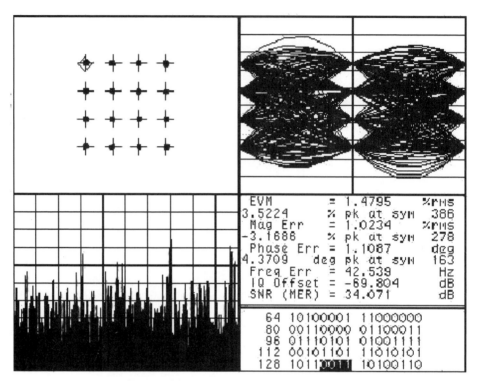

Figure 8.14. Display from a vector signal analyzer measuring a 16-QAM signal.

Modulation error ratio

MER is like a signal-to-noise ratio

EVM is an excellent tool to show the quality of a digital signal for modulation quality purposes, but it does not measure the quality of the signal content relative to the distortion, interference, and noise in the RF environment that these digital signals are subjected to within their channel, namely the channel's composite distortion and noise acting on the digital signal. **Modulation error ratio**, or **MER**, compares the modulation error power, a value similar to the EVM level, to the average transmission power of the channel. The power ratio is measured in dB. Digital channels appear like noise over almost all of the channel bandwidth, so the noise and distortion present in the channel from other sources are difficult to measure, except by turning off the program or measuring an adjacent empty channel. The same difficulty is encountered when making C/N measurement of an analog video signal; the noise is indistinguishable from the signal's program content. MER overcomes this problem by using the signal's own components. An EVM-like error term, the RMS error magnitude, represents the signal's in-channel distortion and noise. The carrier strength is represented by the symbol average magnitude.

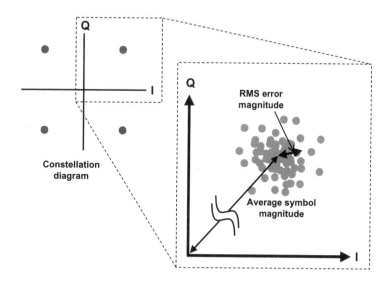

Figure 8.15. Modulation error ratio (MER) defined.

Figure 8.15 shows the relationships. The ratio of these values is a ratio of the signal's strength to its distortion and noise. It compares the digital signal's average power to all the other signal power in the channel. This in-channel power can be noise, distortion, or interference, so that MER is a measure of all the modulation impairments which affect the receiver's ability to recover the data bits. Since it is a power ratio, its units are dB.

MER is the single figure of merit for DVB standards

The measurement DVB Standards Committee for ETS 300-429, adopted MER as the "single figure of merit" for in-channel signal-to-noise performance because it includes all the elements necessary to cover noise and other signal types in-channel. High MER values are good indicators of other system troubles, such as compression and clipping of the digital signal.

The term MER is usually accompanied by the term SNR, for signal-to-noise ratio, because MER is often thought of as the digital equivalent of an analog signal-to-noise test. In a strict definition MER is not signal-to-noise, but its function and purpose are similar. The fact that the MER takes in-channel noise and interference automatically is a real advantage over traditional SNR and C/N measurements.

From a mathematical standpoint, here is how MER is defined:

$$\mathrm{MER(dB)} \; = \; 20 \times \log \frac{\text{RMS error magnitude}}{\text{average symbol magnitude}}$$

The 20 log portion of the equation turns the ratio of the voltage magnitudes into a dB ratio of powers. The number is negative, since the ratio is less than one.

The average vector magnitude represents the power of the signal since it is the average of all the vectors in the signal. The RMS error magnitude is a similar value to EVM. It represents the deviations from the ideal symbol positions caused by noise, distortion and interference. The ratio is less than one since the RMS error is much smaller than the average of the vector magnitude. The closer the ratio is to zero, the higher the number of dB, and the stronger the symbol signal is compared to the interfering in-channel energy. Thus the digital signal is less likely to confuse symbols, which means lower intersymbol interference and better BER.

What is good MER?

Just as with EVM, adequate MER depends upon the system requirements. No measurement is possible if MER is >-18 dB because the average power of the digital signal is unstable. The symbol moves randomly away from its target, and can easily be interpreted by the demodulator as one of the adjacent symbols. For commercial digital modulators, a MER <-30 dB is desirable.

MER diagnostics for troubleshooting and margin

Some system troubles can be diagnosed by observing MER. One is amplifier power compression. MER is quick to respond to amplifier chain power compression because compression causes the average power of the symbol signal to lower while increasing the symbol errors, both effects working to increase the error-to-average ratio dramatically even for small amounts of compression. When MER results are separated into I and Q data, disturbances from noise, carrier leakage, IQ level, and quadrature imbalance can be diagnosed. High MER due to linear phase errors is usually due to carrier frequency error, or spurious interference. Trouble in modulators and some transport impairments are diagnosed by MER information separated into phase and amplitude components. MER can put numbers on the troubles found qualitatively from the constellation diagrams, as seen in "Viewing modulation with constellation and eye diagrams" on page 134.

Margin tells how close the signal is to the cliff edge

But the most valuable use of MER is the observation of the signal's margin. Margin is a measure of how close the signal is to the cliff. Improved MER means better BER. But the relationship is not linear. In fact the relationship looks a lot like the cliff curve for digital signal stability. Figure 8.16 illustrates. Error rate is plotted against MER for the current system. Error rate is better towards the bottom of the graph. MER is better to the right along the horizontal. FEC and adaptive equalization are disabled for the thickest curve on the right.

The margin is the number of dB between the operating point and the point at which the signal would crash. The thin line represents the signal working with FEC and equalization. Because it is closer to the theoretical limit curve, the dotted line, the slope of the curve where margin is calculated is steeper, so the margin is smaller. But the signal can easily tolerate a much poorer MER. The overall effect is that the signal is more robust. Practically speaking, signals used in DVB, such as 64-QAM and higher-bit symbol formats world-wide are never used without FEC and equalization.

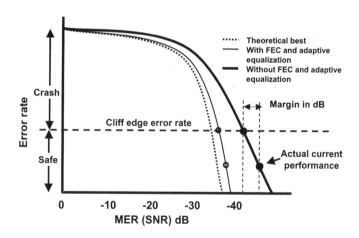

Figure 8.16. MER used to predict the cliff edge margin.

MER predicts margin

From a measurement standpoint such graphs take a long time to create from data streams. To make the margin measurement with out FEC and adaptive equalization operating in the receiver is not practical at all. The margin test is perhaps best used for installation and design operation, although in the future such measurements may be built into simple, cost-effective, and easy-to-use test equipment.

What is good margin?	More is better. Margin is a ratio of MER ratios in dB. If you can record the variations in your system margin over time and environmental conditions, you can set a margin goal that has enough guardband to provide reliable operation through most operating conditions.

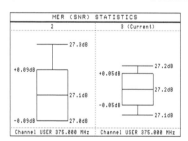

Figure 8.17. Statistical and single MER measurements data displays on a QAM analyzer.

There are many ways to present MER readings. Figure 8.17 show three different displays from a DVB-C QAM analyzer. The top two are single-point readings, updated after a selected number of samples. The bottom display allows the comparison of MER performance between two signals using a more statistical process, in which the minimum, maximum, and mean, as well as a 90% confidence limit set are calculated. With this statistical approach, it is easy to see trends.

Jitter is all in the timing

Modulation quality and the measurement of modulation quality depend upon the system to accurately tell the receiver when a symbol is to occur. Jitter is the uncertainty in the timing of the signal. This concept was introduced in the discussion of viewing digital modulation in "Jitter" on page 87. There are a few, but only a few, remedies available for an RF solution to jitter because most of its sources are beyond your control.

There are two main sources of jitter in a digitally modulated signal: **line jitter** and **peak cell rate jitter**. Line jitter is from the transport system, and is caused by transport line effects such as the phase noise in modulators, the regeneration of digital bit streams, and the power-line coupling to modulators. One source of line jitter is illustrated in Figure 8.11 on page 139, the constellation diagram of residual FM from an unstable local oscillator. Peak cell rate jitter, or PCR, is from the data stream packet and data organization changing cell timing and arrival times. PCR is from multiplexing, compression data servers, or any other reconversion or reprocessing of data from one format or layer to another.

Choosing test points finds jitter sources

If the jitter is originating in the transport line, you have a chance of fixing it. If it is from PCR, then you can only fix it if the signal is being made into packets and/or multiplexed at your head end. Unfortunately both line jitter and PCR look the same on a constellation or modulation measurement analyzer even though origins of each type of jitter are different. Since PCR originates in the digital formatting and reformatting processes, it is much less likely to be seen in a constellation measurement. By selecting the measurement point or points, you can determine its source. Test points before and after remodulation or reconversion of the signal show changes in jitter if its source is line related. Optical fiber transport does not cause jitter. All the conversion to and from RF and lightwaves are rarely, if ever, the source of digital modulation jitter.

Determining clock rate

To measure jitter, the receiver or measuring instrument needs to know the signal's clock rate. Determining the clock rate is done by one of two techniques. On simple, or dedicated receivers, such as set-top boxes, the clock rate for the data stream is built into the hardware, perhaps switched between two or more formats depending upon the set's needs. For more sophisticated receivers, such as cable modems, the clock timing is determined by the receiver itself monitoring the data stream, and reconstructing the rate from the signal, in a feedback arrangement sometimes referred to as clock recovery.

Jitter is observed by using the constellation and eye diagrams on digital modulation analyzers and vector signal analyzers. Measure the digitally modulated signals at the lower system frequencies, because that is where jitter is more likely to affect the signal. If you have access to the channel's baseband signal, use a modulation domain analyzer and frequency interval timer to measure jitter. The measurement capabilities of these instruments are compared in Appendix C, Equipment for Testing Digital Signals in Cable TV Systems.

Instruments for signal quality measurements

Figure 8.18 shows the instrument types associated with each parameter. The parameters, in the dashed-line boxes, characterize the system attributes in the solid-line boxes. Not all of this equipment is available for the measurements at this writing, but as system technologies become standard, these are the products that will probably fulfill installation and maintenance needs.

Measurements on characteristics on the right of the graphs, BER at the receiver, for example, may trigger investigation of one or more paths along the characteristics to the left, using other equipment types, until the sources of the errors are discovered. The bad BER may be traced to modulation quality, which may be traced to frequency and phase response of the system.

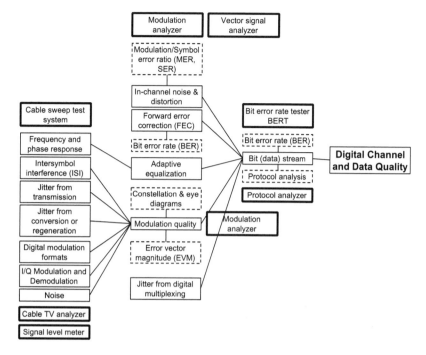

Figure 8.18. Instruments associated with signal quality parameters.

Summary

Measuring signal quality requires the use of qualitative and quantitative techniques, quantitative to measure performance, and qualitative to see where the trouble originates. It is important to put numbers on signal quality to keep track of system changes even when the quality is often hidden by FEC and adaptive equalization. It is also important to determine how close to the cliff your system is operating.

A digital signal can be close to the edge of the cliff, that is, close to failing, without the tell-tale signs of quality degradation an analog signal displays. Bit error rate is the only end-to-end test for most digital signals, but does not tell you how close the system or signal is to the cliff edge.

Just about everything in the path of the signal degrades signal quality. The worst offenders are transport impairments, distortions, and interference. To measure the effects of these, measurements can be divided into three categories: modulation quality, data quality, and the general RF spectrum measurements (power, distortion, noise, and interference). For quantitative measurements, use the constellation and eye diagrams. For more quantitative results, use BER, EVM, MER, and margin.

Bit error rate is an end-to-end system quantitative measure of the data quality of the received bit stream. It is the ratio of the number of bit errors to the total number of bits sent in a given time interval. The cause is always a distortion of the amplitude or phase of the digital signal. Measuring bit error rate usually uses a standard digital stream that the BER tester can check. The measurement usually depends upon the signal and system. Very low BER may take a very long time to measure.

Viewing modulation with constellation and eye diagrams provides good qualitative information for troubleshooting, but does not give measurement data that can be compared against a standard or over time. Constellation diagrams show baseband modulation, and can help impairments, such as noise, spurs, modulator imbalance, compression, and carrier leakage.

Error vector magnitude, EVM, and modulation error ratio, MER, measure signal modulation quality quantitatively. EVM is used to measure the quality of the signal before and after transmission. MER is a comparison of the modulation quality to the noise in the channel. Both can be measured by receivers that imitate the system's set-top box operation. EVM measures modulation quality by assigning numbers to the fuzzy constellation symbol clusters at the signal source and at the receiver for total system signal quality. A measure of the linear distortion of the transmission path can be made by bypassing the FEC at the receiver.

MER is a measure of in-channel signal-to-noise. It derives its signal error information from EVM-like data and its signal power from the average symbol magnitude. It is given as a power ratio in dB. MER measurements are made by digital modulation analyzers dedicated to specific standards and formats such as DVB-C. MER can also diagnose amplifier compression, phase and amplitude error components, and margin. Margin is a measure, in dB, of how close a signal is to the cliff edge. MER takes long time to measure and may not be practical in some systems.

Modulation quality depends upon the frequency stability of the system to accurately tell the receiver when a symbol is to occur. Jitter is the uncertainty in the timing of the signal. Line jitter is from the transport system and is caused by phase noise in modulators, regeneration of digital bit streams, and powerline coupling to modulators. Peak cell rate jitter is from packet and data organization changing cell timing and arrival times. PCR is from multiplexing, servers of compressed data, any reconversion or reprocessing of data from one format or layer to another. Jitter of modulated signals is measured by modulation analyzers and vector signal analyzers. At baseband use a modulation domain analyzer and frequency interval timer. The source of the jitter is determined by choosing the measurement points around them.

Many tools are available to assess digital signal quality. Many of them are dedicated to the specific type of signal and the location of the test point.

Questions for review

1. A good BER indicates proper service delivery. A bad BER indicates impaired service but does not identify the cause of the problem. True or False?

2. Which of these degrade digital signal quality? Choose all that apply.
   a) Transport impairments
   b) Distortion
   c) Interference
   d) All the above

3. Constellation and eye diagrams provide which of the following? Choose all that apply.
   a) BER
   b) Presence of a transport impairment
   c) Baseband data stream
   d) Adjacent channel power
   e) A look at the quality of digital modulation, without putting a number on it.
   f) A graphical display of the symbol pattern. Yes, for the constellation.
   g) A graphical display of the signal trajectory between symbols.
   h) A map used by astronomers and optometrists.

4. BER gives a good indication of how close a digital signal is to the cliff edge. True or False?

5. Error vector magnitude, EVM, does the following. Choose all that apply.
   a) Provides a graph of the symbol position accuracy.

b) Gives a percent representing how close the actual symbol points in a digital modulation signal are to the ideal symbol points.

c) Used for determining how close a signal is to the cliff edge.

d) None of the above.

6. Modulation error ratio, MER, is used for the following. Choose all that apply.

a) A quantitative measure of how close an actual modulated signal is to an ideally modulated signal.

b) A dB power ratio of the total average symbol power to the symbol errors.

c) A value used to determine how close a signal is to the cliff edge.

d) A measure of the in-channel signal-to-noise ratio.

e) All of the above.

7. The source of signal jitter is easily determined by viewing its effect on a constellation or eye diagram. True or False?

Selected Bibliography

1. "Field Installation and Maintenance Testing of DVB Systems," a paper for Hewlett-Packard Company System Engineering Training, Microwave Instruments Division, October 1996.

2. *Using Error Vector Magnitude Measurements to Analyze and Troubleshoot Vector-Modulated Signals,* Hewlett-Packard Company, Product note 89400-14, literature number 5965-2898E, January 1997.

3. Helen Chen, "Testing Digital Video: A Look at Measuring Power and Interference," Communications Technology magazine, Phillips Business Information Inc., Rockville MD, May 1995.

4. Helen Chen, "Testing Digital Video: New Measures of Signal Quality," Communications Technology magazine, Phillips Business Information Inc., Rockville MD, June 1995.

5. Kenneth H. Metz, "Going Digital? Think Bit Error Rate," *Communications Technology* magazine, Communications Technology Publications, Inc., June 1997.

6. David R. Smith, Digital Transmission Systems, ISBN 0-442-00917-8, 2nd Edition, Van Nostrand Reinhold, New York, 1992.

7. Matt Trezise, "Understanding the Measures of Signal Quality in DVB Systems," paper associated with DVB-C 8594Q application training, Hewlett-Packard Company, 1995.

8. Helen Wright, *Digital Modulation in Communications Systems - an Introduction,* Application Note No. 1298, publication number 5965-7160E, Hewlett-Packard Company, 1997.

9. Ian Wright, *HP 8594Q QAM Analyzer Product Note, DVB-C Solutions,* Hewlett-Packard Company, Literature No. 5965-4991E, December 1996.

9

Average Power Measurements

The most critical measurement in your cable or HFC system is power. Digital signal power is a less obvious measurement than an analog signal power measurement because of the signal's nature. This chapter establishes some basics of power measurements before delving into the details of how to make and compare digital signal power measurements.

What you will learn
- The basics of power measurements
- What are RMS, peak, and average power levels?
- What is power bandwidth?
- How is digital signal power measured?
- How to compare analog and digital signal powers
- How is total system power measured?
- Selecting the appropriate test equipment

What is a power measurement?

Power is the energy a signal delivers to a load. Its measure is the most important parameter in your cable television system because it determines whether the signal levels are sufficient for the delivery of information free of interference, distortion, and noise.

Power terms	**Signal power goes by many names, depending upon the signal, system, and specific measurement parameter. A signal power may be called power, level, signal level, carrier level, or strength. Generally, they all mean the same thing: the absolute power of the signal.**

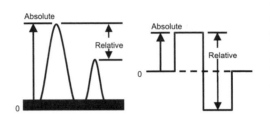

Power is measured as either absolute or relative. **Absolute power** means a value that defines a signal's specific energy level with traceable accuracy. Absolute power is the system design criteria for distribution, such as at the subscriber drop, or as input to RF and optical cable components such as amplifiers. **Relative power** is a ratio of one power to another. It is a differential measurement, with no hint at the absolute power levels involved. The most common applications of relative power levels are distortion, such as CSO and C/N.

Power units review

The high school physics text book definition of 1 Watt is 1 joule/second. A joule is a measure of energy. The per second term means that power is the use of energy over time. Here is a quick review of power terms used in this book that are probably already familiar to you.

In a circuit, a Watt is 1 ampere of current through 1 Ohm of resistance, which produces 1 volt across the resistor. This is Ohm's Law. A dB is a logarithmic ratio of powers, used as a measure of the relative power between two values.

$$dB = 10 \cdot \log\left(\frac{\text{power}}{\text{reference power}}\right)$$

Absolute powers are conveniently measured relative to a constant reference such as a milliwatt (1 mW) or a millivolt (1 mV). These units are dBm and dBmV respectively. Using a voltage as the reference implies a circuit impedance. For example, in North America dBmV is defined to be referenced to 75 Ohms because most cable systems use components and cables with a 75-Ohm impedance.

$$dBm = 10 \cdot \log\left(\frac{\text{power}}{1 \text{ mW}}\right)$$ is an absolute power level as decibel(s) referenced to 1 mW.

$$dBmV = 20 \cdot \log\left(\frac{\text{power}}{1 \text{ mV}}\right)$$ is a decibel referenced to 1 mV in a specified impedance. Even though the reference is a voltage, the value in dBmV is a measure of absolute power, not a ratio, because the units can be traced back to a power in Watts.

The ratio between two absolute powers, in any units, can be expressed in dB. That is, when computing the difference between two dBmV or two dBm numbers, the answer is in dB. For example, if a signals is at -23 dBmV, and a distortion product is at -56 dBmV, the difference in the power ratio is -23 dBmV - (-56 dBmV), or 33 dB. This is commonly stated as the distortion is down 33 dB, or that the two signals are 33 dB apart, or that the relative difference between the signal and its distortion is 33 dB. Because dBm and dBmV are units referenced to constants, they cannot be mixed. For example, it is an error to add or subtract dBm and dBmV units.

The dB provides a convenient numbering unit for discussing power ratios, which tend to be large and ungainly. For example, it is far easier to use +63 dB than it is to use the power ratio $2 \times 10^{+6}$ or -153 dB for the ratio 0.5×10^{-15}.

Power transfer

Power gets work done. At a distance, such as in a cable system, power does work by sending a signal. If there is little loss of the signal energy from one end of the cable to the other, the receiver can use the signal power. In an electrical circuit power is transferred to the user of the power, called a **load**, by a **source** sending an electric **current** at some **voltage**. **Direct currents** (DC) and voltages, that is, currents and voltages that do not vary power over time, do not allow the addition of modulated signals necessary for the transfer of information. **Alternating currents** (AC) and **voltages**, that is, carriers, are effective at sending the messages over long distances through a transmission media such as coaxial cable. Here is why.

An RF cable is modeled as a series of electrical components. Figure 9.1 shows the relationships of inductance and capacitance in a cable medium. This model is called a lumped-element model of the transmission line because the characteristics distributed over the length of the cable are "lumped" into individual electronic inductors and capacitors to show their interaction. **Inductance** characterizes the length along the cable where the energy of the transmission is stored as magnetic fields. Once the current is flowing in an inductor, it resists being stopped. As a consequence, the cable resists changes to the current flowing in the cable, as if the current had physical momentum. **Capacitance** characterizes the relationship between the center conductor and the shield of the cable where the energy is sustained by a charge across the cable's dielectric material. Capacitance resists changes in voltage just as the inductor resists changes in current. This combination of resistance to changes, to current flow along the cable and to voltage across the cable, gives the cable its time-function reaction to changes caused by signal transmissions. The delay in changes to current and voltage suggests that the energy acts like waves, flowing together through the cable in the relatively lossless space between the outer shield and the center conductor. The cable sustains this wave of energy by appearing to the wave as a uniform resistance throughout its distance. This resistance is called **impedance**. Cable impedance is typically 50 or 75 Ohms, depending upon the application. Impedance is not just resistance; it is a combination of resistance, capacitance, and inductance that causes and sustains a phase shift between the alternating current and voltage.

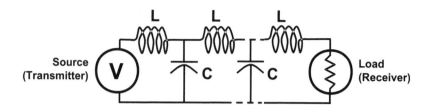

Figure 9.1. Power transfer from a source to a load through electrical elements, inductors (L), and capacitors (C).

Transmission characteristics

The wave-like response of long cables to RF signals is the result of the interaction of the cable's induction along the cable, and capacitance between the center conductor and ground. Induction resists quick changes to current, and capacitance resists quick changes to voltage across the cable. The combination of induction and capacitance gives the cable its characteristic impedance, which is measured in terms of resistance, Ohms, or Ω. Most cable TV cables are 75Ω, although many communication systems are 50Ω.

The impedance and the signals that pass through it are called **complex** because of the addition of the phase relationships between the voltage and current vectors required to measure the signal. Loads at the end of every cable run, such as amplifiers, splitters, and cable connectors, cause changes to the transmission impedance. These variations in impedance appear to the signal as obstructions to smooth power transfer, such as the narrowing of a three-lane road to two lanes. Rather than slowing the traffic down, as happens on the highway, in a transmission line some of the energy gets reflected back. When the electronic highway is uniform impedance, the cable transitions are said to have a good **match**. The traffic has full flow and the power transfer is optimum. A poor match, called a **mismatch**, reflects part of the signal's energy back to the source. The reflections cause amplitude ripples in the frequency response by adding and subtracting with the transmitted signal. The larger the reflection, the wider the peak-to-valley excursions in the ripple.

Digital signals are different

The power of all signals needs accurate measurement

System design and performance are dependent upon the measure, adjustment, and stability of signal power. Signal powers are key to setting adjacent system signals to within acceptable differences, as well as to absolute levels required by system trunk, laser, and other distribution equipment. The amount of distortion in a system is governed, in part, by the total system power. These relationships are shown in Figure 9.2. This chapter teaches the measurement of digital signal average power so it can be compared to its analog neighboring video signals as well as its contribution to the total system power.

Peak envelope power is used as a standard level measure for the analog video carriers of PAL and NTSC standard signals. These signal are tops down amplitude-modulated. This measurement is standardized because it is an excellent and reliable parameter for quantifying power loading in system components as well as for comparing two like-signal levels. The equipment used to measure this very specific type of signal level requires a wide enough bandwidth to capture the peaks of a single video carrier's sync tips without including energy from adjacent carriers.

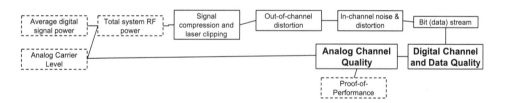

Figure 9.2. Influences of average power measurements.

Digital signal power does not change with signal content changes as does the average channel power of an analog video signal. This is because the digital signal is always carrying a constant-rate data bit stream. The power of a digital signal needs to be measured over the bandwidth of the signal channel, not at the peak of a specific carrier in the channel. But for system design and maintenance to be successful digital and analog signals must be compared accurately and repeatable. Table 9.1 summarizes the analog and digital measurement goals.

Signal Type	Characteristics	Measurement Requirements	Equipment Performance
PAL or NTSC video carrier	vestigial sideband, AM, vertical and horizontal sync pulses at carrier peaks	peak envelope power of carrier	receiver with bandwidth wide enough to measure sync pulse peaks, but narrow enough to exclude adjacent carriers' energy
64-QAM television channel	digital signal in a 6- or 8-MHz bandwidth channel with random noise-like amplitude	average power over entire channel bandwidth	receiver capable of integrating noise-like power over entire channel bandwidth, to the exclusion of adjacent channels

Table 9.1. Matching signal characteristics with measurement requirements and equipment capability.

The following sections teach you how to measure the power of a digital signal, how this measurement compares to analog signal power, and how to set levels to minimize system interferences and distortions. But first, let's review some power concepts.

Types of power measurements

Four types of power measurements

Power is measured according to the characteristics of the signal, the requirements of the measurement standards, and the features and capability of the test equipment. Regardless of these conditions, power measurements can be generalized into four types:

- average
- peak
- pulse or burst
- noise

This chapter covers the absolute and relative measurement of average power and the relative measurement of digital average power to analog video signal peak carrier levels. Peak-to-average, pulse, and noise power measurements are covered in Peak, Peak-to-Average, and Burst Power Measurements, Chapter 10 beginning on page 191, and Chapter 11 beginning on page 211.

Average power

Power transfer of complex RF signals through an impedance still follows Ohm's law even though the computation is more complicated. The power measurement is consistent no matter how complex the phase and amplitude. The goal of the calculation is to find how much a signal would warm a resistor whose value is equivalent to the impedance of the signal's ideal load, that is, a load providing the best match. The standard way to measure this amount of warming is called the **root mean square method**, or **RMS**.

To understand RMS, consider how a signal warms a resistor. Alternating current flows through the load resistor from one direction, then the other, each time causing the resistor to warm a little. The action of cycling voltage and current over the load transfers energy. Power is the product of current and voltage, no matter which way the current flows, or the polarity of the voltage.

Root mean square value (RMS)

The root mean square value, RMS, of a number is a calculation meant to allow comparisons between two dissimilar waveforms. The RMS function takes the waveform values over a time period, squares each linear voltage value, adds them up, then divides out the time units and takes the square root of the result to bring the value back in scale with the original units. The effect is to homogenize the waveform variations to an average or mean value, much like the statistical process of finding a standard deviation of a set of numbers.

In electrical circuits, the RMS value is often called the **effective value** since an AC voltage with a given RMS value has the same warming effect as a DC voltage with the same value. For example, a 10 V_{RMS} AC voltage and a 10 V_{DC} voltage both supply the same power. If the voltage is applied across a 5 Ohm resistor, the power supplied is P = V^2/R = 100/5 = 20 Watts. This is true for any waveform that delivers 10 V_{RMS}.

Voltage times Current equals Power (P=V X I)

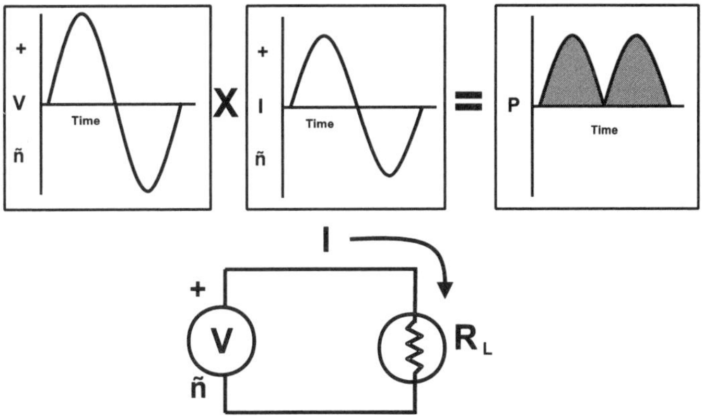

Figure 9.3. Power through a resistor.

In Figure 9.3, the AC power is the product of the current and voltage waveforms. The power is positive, that is, the energy from the voltage source is absorbed by the resistor, because current flowing in either direction causes the resistor to heat up. Even though the "average" of the voltage signal alone is zero, it transfers energy to the resistor.

RMS power is independent of the waveform

To compare the power levels of any two signals over time requires a method independent of the signals' waveform characteristics. In cable television, the RMS or average power of an analog video signal changes with the content of the signal because the modulation format is a form of AM. No one value represents the signal's "average." That is why the peak carrier is used to represent a visual carrier level. The video carrier is by far the largest signal in the NTSC PAL video passband, so it makes sense for it to represent the channel's power.

But how does this help compare the power levels of analog and digital signals? For our purposes average and RMS power represent the total continuous power transmitted in a specified frequency range. **Power** derived from the RMS voltage, V_{RMS}, and RMS current, I_{RMS}, in a constant impedance, is defined as one which repeats its waveform over and over again, that is, it is **periodic**. The power using RMS values is called **average power**, and is given by the following equation:

$$\text{Average power } = I_{RMS} \times V_{RMS} = I_{RMS}^2 \times R = V_{RMS}^2 / R$$

Average power is the power that would warm a resistor, R, if the DC voltage and current were equal to their respective RMS values, regardless of the periodic waveform shape. This is important because it provides a measurement technique to compare the power transfer of any two periodic signals, no matter what their waveforms look like.

Comparing signal powers

Average power, derived from RMS voltages and currents, allows power comparisons between diverse types of signals.

Average power of an AC signal is the net change in energy occurring during one period of the signal. In the case of Figure 9.3, the average power for one period is the area under the power curve, divided by the time elapsed. This is illustrated in Figure 9.4. To get an average power that represents the energy in the time period T, just find the area under the power curve, and divide it by T. The area is in Watts-seconds (Watts times seconds), so dividing by T, in seconds, leaves the average power in Watts. The average power shown as a dashed line below the peak.

<table>
<tr><td>Average power</td><td>Average power is the average of the changes in power over one period of the signal. It can be derived from the RMS values of current and voltage. If the signal does not change power from period to period, its average power is delivered continuously to the load. If the power level varies from period to period, then an average needs to be made over a time period that is long compared to the overall variations.</td></tr>
</table>

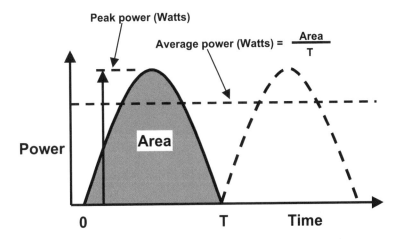

Figure 9.4. Average power calculation of a signal over time.

Data transmissions require a specific bandwidth to transmit information, so the definition of average power must include frequency range as well. The power of such signals is the sum of all the average powers across a signal's frequency span. Combining the concept of average power, illustrated in Figure 9.4 over a frequency band, results in Figure 9.5. In this drawing, the communication signal is shown on the right side of the cube as a signal in the frequency domain. The frequency span is divided up into equal frequency spans shown as slices. The power of each slice's composite signal is shown as a signal in the power versus time dimension of the cube on the left. Each slice contributes to the power display in frequency. The frequency span of each slice is usually a fixed number determined by the measurement equipment and measurement technique.

**Sum power over
a frequency in
volts not dB**

The total power of the signal is the sum of all the slices. From the preceding discussion of RMS power, it is easy to see that the power of each slice must be evaluated in terms of its voltage averages. In instruments which present power in units of dB, such as SLMs and spectrum analyzers, care must be taken to gather or convert the raw power level of each frequency slice average in voltage, not dB, before summing up. Often the scales of the measurement instrument can be changed from dB to volts by switching the scale from logarithmic, or log, to linear.

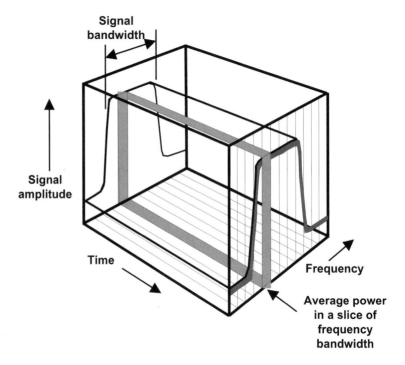

Figure 9.5. Average power over a frequency band.

Power and bandwidth

The power measurement of signals with no visible or prominent carriers requires the use of averaging techniques over time and frequency, such as shown in Figure 9.5. The idea that a signal's power is distributed over a finite frequency leads to the definition of a power over a frequency band, called power bandwidth. **Power bandwidth** is the total average power of a signal contained in a specified frequency range. There are different ways to specify power bandwidth.

- The absolute power contained in a bandwidth specified by the shape of the signal in the frequency domain, in dBmV, dBm, or any other absolute power unit
- The percent of total signal power contained in a specified 3-dB bandwidth, in percent (%)
- The bandwidth of a signal containing a given percent of the signal's total power, in Hz for a percent of total power
- The absolute power per Hz at a given frequency, in dBmV/Hz, dBm/Hz, or any other absolute power density unit

---

Power bandwidths

The measure of power bandwidth is only useful for signals whose power stays constant over time within the bandwidth. For analog PAL or NTSC television signals, whose power levels change according to the signal content, power bandwidth is not a meaningful measurement because the peak power within the analog channel's 6 to 8 MHz bandwidth is not a constant. But for a digital signal, whose modulation techniques spread information power over most of the allocated bandwidth, power bandwidth is an important parameter.

Table 9.2 shows graphic representations of these power and bandwidth measurement definitions. Average signal power and noise power are used in cable television systems. Power bandwidth, percent power bandwidth, and noise power are common in wireless communications systems for mobile and cellular radio.

Bandwidth defined

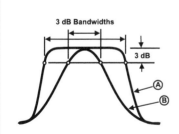

3 dB Bandwidths

3 dB

Ⓐ
Ⓑ

Systems, filters, and signals have bandwidths. The term means a range of frequency over which the system, filter, or signal operates. For a system, bandwidth is the operating range, such as the 450=MHz bandwidth of a cable television system. Bandwidth does not tell the absolute frequencies, just the range covered. Bandwidth usually implies a frequency response curve that is expressed graphically, like the drawing on the left. Curves A and B can represent the response of either a filter or a signal. Their bandwidth shapes are usually defined relative to the highest response points, shown here at the responses' center frequency. Trace the curve on either response to a point where it decreases 3 dB, mark the point, and repeat on the other side of the center frequency. This frequency span between the "3-dB points" is the 3-dB bandwidth of the filter or signal in Hz.

The 3-dB roll-off is not the only bandwidth reference used. The 3-dB points are most commonly used because they represent half of the power of the maximum response reference (remembering that 3 dB equals 10 x log 1/2, where the 1/2 is the power ratio). For a digital signal the 3-dB bandwidth can be interpreted as the symbol rate. Sometimes 6-dB points are used, representing 1/4 of the power at each roll-off point. Signal or filter bandwidths cannot be directly compared if they use different roll-off amplitudes, but they can be corrected.

In each definition, the bandwidth of the signal is referenced. Channel bandwidth is specified as a frequency span, such as 8 MHz for a PAL television channel. For analog signals and many digital signals the bandwidth is also the **channel spacing**. Power bandwidth shapes and their defining parameters vary with signal format and measurement techniques, as illustrated in Figure 9.6. Digital signals, unlike analog video, tend to be symmetrical about the center of their channel.

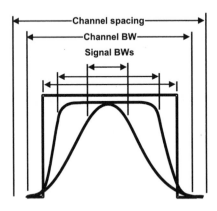

Figure 9.6. Terms defining various bandwidths.

The shapes of the signals in Figure 9.6 are determined by the communications signal specifications, the square shape being ideal, but not very realistic. The other, softer sided shapes keep the signal energy confined more easily to the channel boundaries. These frequency domain shapes are generated by the time domain pulses that make up the digital signal. From "What is in a signal?" on page 25, this shape comes from the time domain pulse shapes that are rounded to conserve frequency spectrum and keep spurious signals to a minimum. For a digitally modulated signal the bandwidth containing more than 99% of the signal's power defines the reciprocal of the signal's symbol rate.

Specification name	Appearance	Required parameters	Use and example
Average signal power		Power algorithm (peak or average), for specific types of digital modulation, the signal bandwidth is assumed.	Average channel power of a digital signal such as a QAM or VSB video signal, measured over specified signal bandwidth, in dBmV, dBm, or other absolute power units.
Power bandwidth		Channel or signal bandwidth	Bandwidth in Hz containing 99.9% of the signal power. Used in wireless transmitter performance for digital signals.
Percent of total power in bandwidth		Channel or signal bandwidth	Percentage of total signal power in a given bandwidth. Used for wireless transmitter performance.
Noise Power		Reference bandwidth for power density, correction factors for measurement instrument	Broadband noise power, system noise of cable systems for computation of carrier-to-noise ratios, in dBmV/Hz, dBmV/4 MHz, or dBm/Hz.

Table 9.2. Absolute power measurement techniques and terms.

Signal level meters and spectrum analyzers

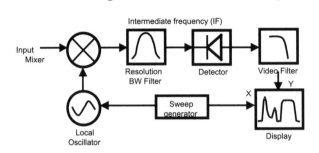

Signal level meters (SLM) and spectrum analyzers share the same radio receiver block diagram shown here. Here is how this works: the LO frequency mixes with the incoming signal to produce a difference frequency, called an intermediate frequency (IF) signal that is processed to provide the test results for display. The detector converts the IF signal to a very low frequency signal while retaining the signal's amplitude information. The video filter is a low-pass filter that smooths out jumpy noise responses for easier interpretation. The resulting amplitude is displayed on the screen as a point along a frequency trace, or as an amplitude reading. This heterodyne tuning scheme enables the test instrument to cover a wide frequency range, while the IF circuits, only having to work at one comparatively low frequency and narrow bandwidth, focus on accurate amplitude and frequency performance.

The IF filter determines the shape of the input signal in the test equipment's display. It is like an RF window through which the instrument input is viewed. If the window is narrower than the signal, the power measured is less than the signal's average power. If the filter is wider than the signal's bandwidth, the total power of that signal can be measured, but there is the chance that adjacent signal power may add inappropriately to the power reading. Spectrum analyzers and SLMs generally do not have IF bandwidths wide enough to measure a digital video channel in one reading. Rather, they have internal software procedures that measure the signal's power by summing up the individual power readings while the instrument tunes through the digital signal, as illustrated in Figure 9.5 on page 164.

Relative signal level measurements

The most important relative measurement is the adjustment of adjacent channel levels. So much of system performance depends upon the accurate and consistent relative powers of system signals that this measurement needs to be routine, just as it is in today's analog systems. Changes in the signal levels can indicate suckouts and roll-off due to equipment failure or misadjustment. System frequency response has a dramatic effect on the system's digital signals.

There are two important considerations for the relative powers:

- How are digital signal powers measured?
- What is the optimum relative power level between an analog and digital signal?

Adjacent channel power compares signal levels

The measurement of digital signal absolute powers has been automated. Making manual power measurements of digital signals is possible, but tedious. Automated tests measure **adjacent** channel powers. The common name for these are **adjacent channel power** measurements, or **ACP**. Comparisons to channels further away in the band are called **alternate** channel powers. These tests are summarized in Table 9.3.

Measurement		Critical parameters	Definition
Adjacent channel power (ACP)		Channel bandwidth and spacing	Compare adjacent signal levels, or, if the adjacent channel is empty, to measure the power of the main signal in the adjacent channels. Use head end modulator to ensure a good start with no initial adjacent interference.
Adjacent channel leakage (ACL)		Channel bandwidth and spacing	Leakage in adjacent channels may be a symptom of digital modulator problems: faults in the modulator output filter, or overdriven modulator. Tests can be done in-service from a modulator test port or directional coupler off the feed.
Alternate channel Power		Channel bandwidth and spacing, spacing of the channels beyond the adjacent channels	Used to search for interference and spurs in unoccupied channels. Primary use is for wireless transmitter out-of-channel distortion tests.
Peak-to-average power		Signal peak power and average signal power	Signals with high peak-to-average ratios subject system components, such as optical amplifiers to overload and clipping.

Table 9.3. Relative power measurement techniques and terms.

The answer to the second question is provided by the experience of systems in actual practice. The analog signals are measured by following the same rules as before, using the carrier peak level. Procedures later in this chapter show how to measure digital signal average power.

Adjacent channel power not only affects the adjacent channels, but the conditions that cause the adjacent channel power spills usually affect the modulation quality of the primary signal.

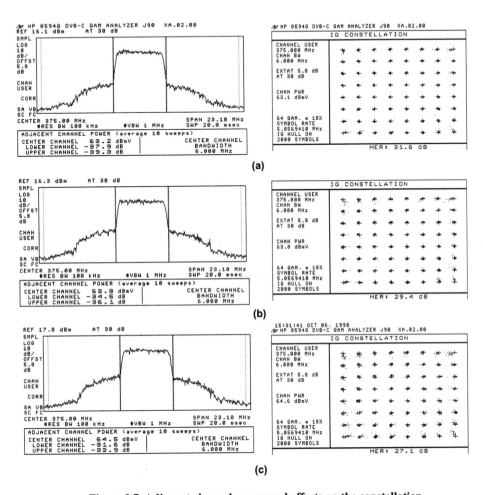

(a)

(b)

(c)

Figure 9.7. Adjacent channel power and effects on the constellation.

Digital signals are less likely to be interfered with by analog signals than vice versa

In Figure 9.7, a 64-QAM signal is distorted by an amplifier at three different levels. In (a) the left screen shows the distortion and power readings. The constellation and MER do not seem troubled. As overloading increases, in (b) and (c), the adjacent channels get more of the third-order products and the constellations show the compression of the amplifier.

The robust nature of digital signals means they are more likely to cause interference in neighboring analog signals than the other way around. A digital signal produces random spikes of energy that occur during a specific set of digital signal sequences, like the occasional sparks that fly off a smoothly burning fire. These spikes are called digital signal peaks. If the digital signal level is set too high compared to adjacent analog signals the peaks can cause interference with the lower adjacent analog channel sound, or the upper adjacent channel video. The level of digital signal peaks is measured relative to the average power of the signal, called peak-to-average ratio, discussed in the next chapter.

Power is measured using different procedures

To prevent interference with analog signals, a digital signal is typically set as much as 12 dB lower in level than an adjacent analog signal. The differences are based upon measuring the analog signal as a peak response, and the digital power as average total channel power. Actual system test evidence, where the analog and digital signal groups were restricted to their own frequency ranges, demonstrates that digital signals 6 dB to 10 dB below the analog signals resulted in low internal distortion and good error performance for the system. In the experiments, the 6 dB difference seemed to be optimum best for both analog and digital quality. More on distortion is found in Distortion, Noise, and Interference, Chapter 11 beginning on page 211.

Adjacent digital and analog signal levels	Digital channels should be 6 to 10 dB lower than analog signals whether adjacent or in different frequency bands, to keep interference between them minimized. The analog signal is measured as a peak level (as for analog performance) and the digital signal power is an average channel power measurement.

Measuring digital signal average power

Maintaining the proper average transmission power is a key adjustment made in cable systems for digital as well as analog signals. Unlike the narrowband measurement of the analog visual carrier level, testing average power of a digital transmission is a wide bandwidth measurement. The instruments that can make the various average power measurements are shown in Figure 9.8. There are a number of choices for each measurement parameter. This section has procedures to help you decide which instrumentation is best for your needs.

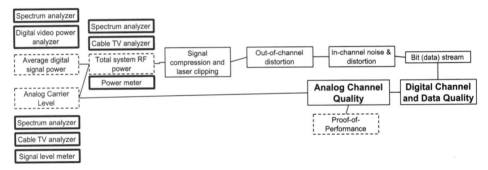

Figure 9.8. Performance and measurement map with instruments that measure average power.

Use a channel bandpass filter with a power meter

One method of measuring average digital video power is to use a **filter** and a **power meter**. This is illustrated in Figure 9.9. The bandpass filter must be flat across the entire channel bandwidth, 6 to 8 MHz wide depending upon your system. The power meter's sensor, the part that detects the RF energy, is usually a much wider frequency range than the channel; the filter is necessary to prevent measurement of other channels' power. It must be centered on the signal, and provide a sharp roll-off to remove energy from adjacent channels. Rejection, the attenuation a filter provides outside its passband, must be between 50 to 60 dB. The passband ripple, similar to the peak-to-valley specification for in-channel frequency response, must be less than ±0.5 dB to maintain the power meter's measurement accuracy.

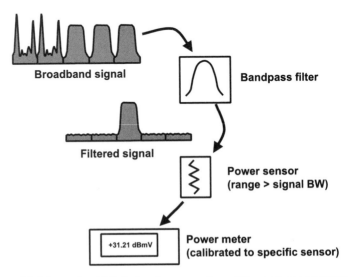

Figure 9.9. Measuring digital signal power using a filter and power meter.

The digital video signal is applied to the filter and power sensor, reading the true average power. This method is accurate and inexpensive for a signal channel test. Note that the power meter must have sufficient bandwidth and frequency range to handle 6 to 8 MHz wide signals at up to the highest carrier in the cable system, optimally up to a 1 GHz carrier frequency. One filter is required for each channel to be tested, unless a tunable channel filter is available. Unfortunately, tunable filters generally do not provide sufficient rejection out-of-band over their entire frequency range.

Tunable filters must be fully characterized for noise power bandwidth versus frequency, because even the slightest change in bandwidth makes a large noise power difference. See the lowest row of Table 9.4 on page 177. If possible, measure the absolute power of one channel with a well-characterized fixed-tuned channel filter, then measure it again without the filter, but with the same analyzer settings. Now set the other channel levels in the group relative to the first channel's unfiltered level.

The power meter and filter provide the best average power level accuracy because the power meter and its power sensor are designed for just that purpose: measuring the total power of a signal by how much it heats the load. Because of its accuracy the power meter is often used at the head end to provide exact launch levels. And, because the individual channels are isolated, a filter is not required, reducing the measurement uncertainty even more.

Power meter is most accurate

For measuring the power of a digital signal, the power meter is more accurate, but more awkward to use in all but permanent monitoring installations.

SLM is designed to measure specific types of digital signals

A second technique uses the familiar **signal level meter**, which has been adapted to make digital signal power measurements. Although not as accurate as a power meter, the signal level meter, or SLM, is very portable, and capable of both analog and digital signal tests. The SLM is the choice for field measurements. With the proper built-in algorithms, the signal level meter does not need any external filtering to make accurate power measurements.

The third way to measure average power uses a **spectrum analyzer**, **signal level meter (SLM)**, or **cable TV analyzer**. Since these analyzers do not have a receiver IF bandwidth wide enough to measure

Use built-in test procedures in spectrum analyzer and SLM

the whole digital signal at once, they sweep through the signal, sampling and summing data for the average power computation, as shown in Figure 9.5 on page 164. Since most spectrum and cable TV analyzers are designed for a wide variety of signal and noise measurements, they may not have traditional square-sided, flat-topped filters used to measure these noise-like signals, so the calculated average power is corrected for the shape of the filters actually used in the analyzer. These filter definitions are called the analyzer's **noise equivalent bandwidths**. For a complete discussion and analysis of the accuracy of a spectrum analyzer in making noise measurements see Gorin in the bibliography at the end of this chapter.

Spectrum analyzer precautions

The spectrum analyzer is a wide-bandwidth receiver with a very power sensitive input circuit. For this reason damage levels can occur at +59 dBmV (+10 dBm) at the analyzer's input mixer. This power does not have to be all in one signal, it can be spread over the entire analyzer input frequency range to affect the analyzer's performance.

A manual measurement would require a measure of the average power at each of the up to several hundred points, summing these using an RMS technique. Fortunately, most spectrum and cable TV analyzers and signal level meters have built-in power tests that do these tasks automatically.

Information you need to make automatic channel power tests

Testing with an SLM, cable TV analyzer, or spectrum analyzer requires that you know some specific attributes of the signal, and are able to set the test equipment parameters listed below. Many of these settings have to be set manually, even when the measurement procedure is continued automatically by the instrument.

- Channel spacing in MHz
- Signal is at least 10 dB out of the analyzer's noise
- 3-dB channel bandwidth in MHz
- Resolution bandwidth setting in kHz (spectrum analyzer only)
- Video bandwidth setting in kHz (spectrum analyzer only)
- Minimum number of signal averages
- Level correction for analyzer bandwidth shapes, in dB

To guarantee the accuracy and the repeatability of the power measurement, the channel under test should have a maximum response that is at least 10 dB out of the analyzer's noise. If the noise is less than 10 dB, the measurement will be inaccurate and unrepeatable because the analyzer's noise combines with the input system noise. And, because the analyzer is subject to overload due to its wide frequency range, the noise viewed on the analyzer's display may be caused by distortion in the analyzer. Here is a quick and easy procedure to check both the signal-to-noise ratio (or noise-to-noise ratio in this case) and input overload distortion.

Checking the analyzer's signal-to-noise ratio

1. To assure that your measurement does not include the analyzer's noise, set a marker on the top of the carrier, and a delta marker on top of that one, as shown.

2. Reduce the video bandwidth to 300 kHz.

3. Disconnect the analyzer's input signal.

4. Read the drop as the marker difference in dB. If this value is less than 10 dB further power measurements are going to be inaccurate. This example shows a 21.4 dB ratio assuring that the measurement is accurate.

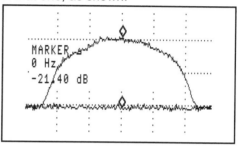

5. To increase the signal-to-analyzer-noise ratio, reduce the analyzer's input attenuation value. A zero dB setting may be necessary.

6. Recheck the signal-to-noise ratio.

7. With the signal connected decrease the analyzer's attenuator by 10 dB and look for a decrease in the noise floor of 10 dB. Any other noise change than 10 dB indicates that the analyzer is overloaded. Use the highest setting of the attenuator which causes a 10 dB change in noise.

8. Retest the analyzer for sufficient signal-to-noise ratio.

Channel bandwidth is the shape of the power amplitude over its frequency range. The power bandwidth shape affects the measurement of total average power whenever the measurement equipment is used to estimate the total power from samples rather than taking data over the whole channel. Following is a simple procedure for estimating the channel bandwidth for use in estimating the signal's average power using manual marker methods.

Measuring the 3-dB channel power bandwidth

1. Center the channel under measurement in the spectrum analyzer display and set the analyzer span equal to the channel bandwidth.

2. Narrow the video bandwidth to reduce the ragged top of the signal as shown. In this example the video bandwidth is changed from 30 kHz to 100 Hz.

3. Increase the analyzer's resolution bandwidth to allow the digital signal's digital energy to fill in the display. As you increase the bandwidth note any changes in the shape of the signal. When the signal pattern begins to flatten, then it is the analyzer's resolution bandwidth being displayed, not the signal shape. Back down the RBW setting by one level. In this case a 30 kHz RBW is used.

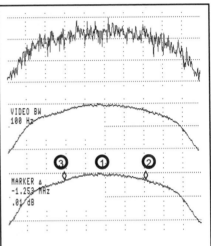

4. To measure the 3-dB power bandwidth of this signal, place a single marker at or near the center or the peak of the signal. If the display is moving so much that the marker reading varies more than a few tenths of a dB when sitting at a single position, smooth the display further by narrowing the video filter, or turning on video averaging function, then stop the trace from being updated. This is often called the view mode.

5. Use peak search to find the largest amplitude response. Call up a second, or delta, marker at this point, and move this marker off to the left or right until the differential amplitude reads 3 dB.

6. The channel's power bandwidth is twice the differential frequency reading in the marker's display.

7. If the channel response is too flat to find the center accurately, place the marker, (1), to a point representative of the flat-top level. Use delta to move the second marker, (2), to one side so that the amplitude difference between the two markers is -3 dB. Press delta again to place a new second marker on top of the first adjusted marker. Now place the new marker, (3), on the opposite side of the signal response so that the amplitude difference between markers reads 0 dB. The frequency readout is equal to the the signal's power bandwidth.

Average power measurements require strict adherence to measurement procedures because there are many parameters associated with digital modulation formats. Among the techniques available with spectrum analyzers, some are much easier than others. Table 9.4 gives a summary of the procedure types with spectrum analyzers and the associated ease and accuracy of the measurements you can expect.

Spectrum analyzer procedure	Ease of measurement	Accuracy
HP 85721A Cable TV Measurement for HP 8591C and DVB-C HP 8594Q QAM Analyzer	Excellent	Excellent - measures entire channel automatically
HP 8590E Series	Very good	Excellent - measures entire channel automatically
Noise marker	Difficult	Good - requires calculating analyzer's 3-dB bandwidth shape
Manual	Very difficult	Poor - requires manually calculating shape around the peak response

Table 9.4. Procedures for making average channel power measurements.

Here are the settings suggested for two types of signals using the spectrum analyzer with a cable television channel measurement and a general analyzer built-in average channel power routine.

Measuring 64-QAM average channel power

1. Measure a continuous 64-QAM signal having a 5-MHz bandwidth.

2. With a Hewlett-Packard 8591C spectrum analyzer with the HP 85721A Cable TV Measurement personality, use the Power sub-program with channel bandwidth set to 6 or 8 MHz depending upon the system channel and signal bandwidths. For the built-in routine, manually set the channel bandwidth to 6 or 8 MHz. Channel spacing also needs to be set according to the system requirements.

3. The HP 85721A program automatically sets the resolution bandwidth (RBW) to 100 kHz. For the built-in routine, you need to set the RBW to 100 kHz manually. This RBW assures that each of the frequency points measured along the channel bandwidth has time response to gather the noise-like digital information, but narrow enough to exclude the energy of adjacent frequency slots.

4. The program automatically sets video bandwith (VBW) to 1 MHz. This wide bandwidth allows all the time-varying signals to be detected. A narrow video bandwidth would exclude the higher-frequency components of the digital signal.

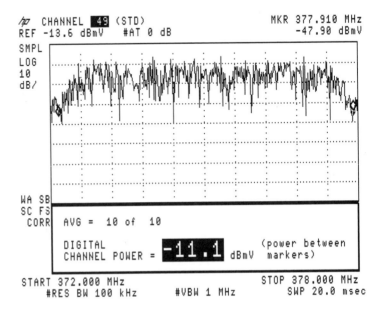

Figure 9.10. 64-QAM signal channel power measurement.

5. Measure three consecutive times and average these values by adding the powers in dBmV and dividing by 3 to further reduce variations in the time collection of data. Note that this mathematical procedure is an approximation used to save time and effort. Since the values already represent averages, the error in manipulating dB instead of μWatts, is very small.

6. For either the software or the built-in routine set the average count to 10 to get an average value quickly. This assures that the measurement represents the signal's overall energy content. Figure 9.10 shows the power reading from the software measurement.

7. If you are using the built-in measurement routine, add 2.5 dB to the power to get the 64-QAM RMS signal power. This value accounts for the linear averaging of the logarithmic signal data. For example, if the power displayed by the routine is -13.6 dBmv, adding 2.5 dB gives -11.1 dBmV as the corrected power of the signal.

Other manufacturers' spectrum analyzers may have slightly different average power subprograms, so be sure to use their guidelines for proper correction values and accuracy tolerances.

Setting spectrum analyzer bandwidths for digital power measurements

Not all procedures in spectrum analyzers automatically set the resolution and video bandwidths for accurate measurement results. The resolution bandwidth of a spectrum analyzer must be set relative to the bandwidth of the digital signal being measured to account for a digital signal's shape. If the analyzer's bandwidth is set too wide, the roll-off of the signal band edges are read incorrectly, resulting in large errors. If the analyzer's bandwidth is set too narrow, the filter cannot respond to the high frequencies of the digital signal, resulting in low power readings. Here is the general rule to follow when setting the analyzer's resolution bandwidth.

$$\frac{\text{Channel 3-dB bandwidth}}{20 \text{ to } 100} = \text{Resolution bandwidth}$$

For example, if the 3-dB bandwidth is 5 MHz, the resolution bandwidth can be between 250 kHz and 50 kHz. From a practical standpoint, simply pick the bandwidth between these limits available on you analyzer, such as, 100 kHz. The procedure for measuring bandwidth is in "Measuring the 3-dB channel power bandwidth" on page 176. The video bandwidth should be at least three times the resolution bandwidth to prevent filtering of the digital signal responses.

A spectrum analyzer is designed for the measurement of CW signals and broadband random noise. Since digital signals have some unique signal attributes, each procedure requires its own correction factor.

Measuring QPSK average channel power

1. This procedure is for a QPSK signal whose channel spacing is 3 MHz.
2. Using the software channel power program with the channel spacing set to 3 MHz. The program sets the RBW to 30 kHz and the VBW to 100 kHz. These settings keep the frequency slot information accurate for the narrower bandwidth. This is illustrated in Figure 9.11.

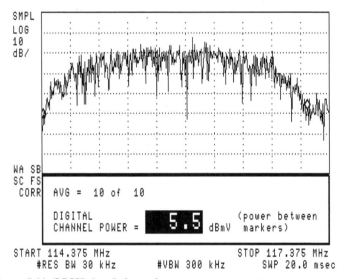

Figure 9.11. QPSK signal channel power measurement.

3. The power level, shown between the diamond-shaped markers, is +5.5 dBmV.

A DVB-C QAM signal power is shown in Figure 9.12 made by a fully automatic average power measurement in a spectrum analyzer dedicated to that format. Power readings in the portion of the trace between the two vertical lines at the bottom of the screen are used to obtain a true power average as in the QPSK measurement of Figure 9.11. Instrument software optimizes the accuracy of the measurement by adjusting the spectrum analyzer settings and applying the internal filter amplitude correction. This simple method of measurement is used on channels across the full frequency range of the spectrum analyzer and over any desired bandwidth, eliminating the need for external channel filters. The accuracy of the power reading can be optimized to less than ±1.0 dB, not as accurate as an SLM or power meter, but much more versatile in measurement flexibility.

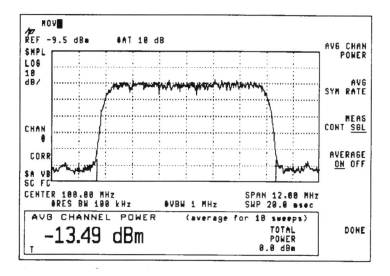

Figure 9.12. An automatic digital signal power measurement on an HP 8594Q QAM Analyzer.

Channel power using a noise marker

The noise marker in spectrum and cable TV analyzers assists in making broadband system noise measurements by automatically applying corrections for the analyzer's resolution bandwidth, shape, and the detection scheme. A digital signal is noise-like, but it is confined to a channel bandwidth, not broadband. Since the noise marker assumes a broadband noise, its answer needs correction for the specific digital channel bandwidth being tested.

Measurements on channels that are less of a square-top shape are much less accurate. Comparing the differences between the power of channels with similar shapes can be made with great accuracy due to the excellent repeatability of most spectrum analyzers. Here are the steps necessary to measure the absolute channel power of a digital channel with a spectrum analyzer noise marker.

Measuring channel power using the spectrum analyzer noise marker

1. Find the analyzer's noise marker function, usually under a key or softkey in the marker functions. Turn the noise marker on.
2. Set the span to view one channel in the display plus about 20%. That is, set the span to equal the channel bandwidth x 1.2. (Although the next figure shows a 6 MHz span.)

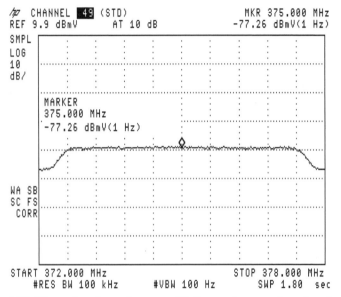

Figure 9.13. Measuring channel power with the spectrum analyzer's noise marker.

3. Set the RBW of the analyzer to the closest value shown in the equation of "Setting spectrum analyzer bandwidths for digital power measurements" on page 179. For example, for a channel bandwidth of 5 MHz, the RBW must lie in the range 5 MHz/20 to 5 MHz/100, or 250 kHz to 50 kHz. The analyzer RBW that fits this range in one whose standard RBW sequence is value ratios of 1, 3, 10, is 100 kHz. Set the RBW to 100 kHz.
4. Set the VBW to a value equal to or less than 1/1000th of the RBW. This sets the measurement speed to provide a stable readout value. For this example set the VBW to 100 kHz/1000, or 100 Hz.
5. Find the 3-dB bandwidth of the channel power by using the procedures in "Measuring the 3-dB channel power bandwidth" on page 176. The 3-dB bandwidth is 5 MHz.

6. Turn on the noise marker and place it near the center of the channel response and read the level in dBmV/Hz. This is illustrated in Figure 9.13. The noise marker level is -77.26 dBmV (1 Hz). The 1 Hz reference means the noise power in a 1 Hz bandwidth.

7. Multiply the power level by the power bandwidth in Hz to get the digital signal power. The power level in a 5 MHz bandwidth is 10 log (5 MHz/1 Hz), or 67 dB. Add the 67 dB to the value of the noise marker to get the noise power over the whole bandwidth. The signal power is -77.26 + 67 = -10.26 dBmV.

A spectrum analyzer without built-in power functions or a noise marker presents a challenge, but here is an overview of the necessary steps.

Measuring digital channel power manually with a spectrum analyzer

1. Set the analyzer's detection mode to **sample detection**. The default detection mode for most spectrum analyzers is peak to catch the most powerful response in any one frequency slot. To insure that the display represents a noise-like random signal, you must select the detection sample mode. With sample detection a random value is placed in each frequency slot. Peak detection skews the reading to a value higher than actual.

2. Set the RBW of the analyzer to the closest value shown in the equation of "Setting spectrum analyzer bandwidths for digital power measurements" on page 179. This example uses 100 kHz. Set the VBW to a value 1/1000th of the RBW and set the video averaging function until the signal level converges to a line. VBW is set to 100 Hz.

3. Caution: Don't be tempted to save time by increasing the VBW to get faster trace updates. Your power readings will inaccurate and unrepeatable.

4. Activate the marker and place it at the middle of the channel.

5. Find the 3-dB bandwidth of the channel power by using the procedures in "Measuring the 3-dB channel power bandwidth" on page 176.

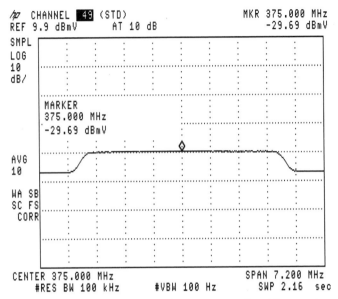

Figure 9.14. Measuring signal channel power manually with a spectrum analyzer.

6. The correction factor for the RBW filter and log detection is +
   2 dB for 3-dB bandwidth filters, and 3.6 dB for 6-dB bandwidth
   filters.
7. Correcting the power at one RBW point for the channel
   bandwidth is 10 log(ratio of bandwidths). For a 100 kHz RBW
   and a 5 MHz channel bandwidth this correction is 10 log(5
   MHz/100 kHz), or +16.99 dB.
8. Apply both correction factors to get the channel power: -29.69
   dBmV + 2 + 16.99 = -10.69 dBmV

In selecting equipment for these measurements, you need to consider both accuracy and utility, summarized in Table 9.5 and Table 9.6.

Measurement instrument	Accuracy (typical)	Advantages	Disadvantages
Power meter and band pass filter	±0.3 dB	Simple procedure, accurate, inexpensive, results less dependent upon measurement process, portable	Requires a fully characterized bandpass filter for each signal. Filter needs 60 dB rejection, and flat passband response
Signal level meter	±0.5 dB	Simple procedure, can have automatic testing built in, very portable, no external filters	Cannot view signal while measurement is being made
Spectrum analyzer	±1.0 dB	Measure any signal in system, can have automatic testing built in, no external filters	Manual measurements are less accurate due to the various correction factor requirements. Because of the calculations, they also take more time and are subject to operator errors.

Table 9.5. Measurement techniques for absolute average digital signal power

Equipment for testing digital signal average power	Primary use	Secondary use
Power meter and filter	Installed power monitor at head end or hub. Device manufacturing.	Quick and simple field measurement
Signal level meter	Portable field use for digital and analog test and maintenance	Quick check at test points in and around head end and hub
Spectrum analyzer	Monitor wide variety of head end input and output	Performance verification throughout system

Table 9.6. Where to use what test equipment for digital signal average power.

Total system power

In HFC systems laser clipping sets total power limits

The limitations to power transfer in the forward and return direction of a hybrid/fiber system is the total system power to be sustained by the optical fiber network. Clipping of the signal in the fiber path seems to be the most common first-level overload condition of HFC systems. Clipping shows up in the cable system as excess distortion, even when all the signal levels seem to be well-behaved. Although it is not a day-to-day measurement, determining total system power is critical for installation and turn-on. Some spectrum and modulation analyzers automatically determine the total power input to the instrument before preceding with a measurement. Caution must be used when relying on these automatic tests; the operating instructions for entering channel parameters may not tell the analyzers' internal software enough about the signals to assure accurate analyzer attenuator/preamplifier settings. For a discussion on distortion topics, see Distortion, Noise, and Interference, Chapter 11 beginning on page 211.

The computation of system total power for analog systems is well understood. NTSC and PAL signals' average powers are estimated from their peak carrier levels. Digital signal average powers add directly to the cable total. No estimations or corrections are necessary. Summing the powers of these channels is treated the same way a digital signal power is summed, by converting each channel power to Watts, and adding the powers, then converting the Watts back to dBmV or dBm.

An analog television signal's average power is less than the power of the peak carrier. So the sum of all the analog carriers in a system is less than the sum of the peak powers. The visual carrier level is a predominant part of the power in the television channel. Accompanying audio carriers in the channel are 7 to 13 dB below the visual carrier so their power contribution is only 1/5th to 1/20th of the peak of the visual carrier. Since these levels are within the expected uncertainty of the measurement, they are ignored for estimating purposes. The peak of the video carrier is 2.5 dB higher than the black level of the horizontal information. As a rule-of-thumb, the average power of an analog television channel is about 3 to 4 dB below the carrier peak level, as measured for performance testing. If in doubt, use the average power measurement procedures in this chapter to substantiate the average power of the analog signals.

Here is an example of calculating total system power.

Estimating total system power

1. The system is 75-Ohms impedance with 2 +8 dBmV pilots, 25 +5 dBmV analog PAL channels, and 30 -6 dBmV QAM channels. Estimate the total power in the system.
2. Calculate the power contribution of two +8 dBmV pilots: each +8 dBmV = 84.1×10^{-9} Watts, or 0.0841 µWatts. For two pilots the total is 0.168 µWatts.
3. The contribution of the analog channels: each +5 dBmV = 0.042 µWatts. For 25 channels, the power is 25 x 42 µWatts = 1.050 µWatts.
4. The contribution of the digital channels: each -6 dBmV = 0.0033 µWatts. For 30 channels the power is 30 x 3.3 µWatts = 0.099 µWatts.

5. Total system average power = 0.168 + 1.050 +0.099 = 1.317 μWatts.
6. Total system average power in dBmV = +19.97 dBmV.

The digital signals are 11 dB below the analog signals, so their contribution to the total power of the system is very small. Even if the digital signals were 6 dB lower, their contribution would be small.

Estimating total system power	The total average power of an analog television channel is approximately 3 dB less than the peak carrier level. Digital channel power is measured by using the average power procedure appropriate for its modulation and channel bandwidth. All powers are converted to Watts before summing them. Convert the power in Watts back to dBmV.

For more examples on unit conversion, see Thomas in "Selected bibliography" on page 189.

Summary

Power is the energy a signal delivers and its measure is the most important parameter in your cable television system. Power is measured as either absolute or relative. An RF cable is modeled as a series of electrical components which act as a complex impedance transmission line that influences the way signal power is transferred.

System design and performance are dependent upon the measure, adjustment, and stability of signal power. The analog signal for the top-down amplitude-modulated video carriers used for PAL- and NTSC-standard video carriers is a measure of the carrier peak envelope power. Digital signal power does not change as the signal content changes. The opposite is true for an analog video signal. There are four types of power measurements: average, peak, impulse, and noise.

Average and RMS power, the power it takes to warm a resistive load, is the total continuous power transmitted in a specified frequency range. Data transmissions require a specific bandwidth to transmit information, so the definition of average power must include frequency range as well. The total power of the signal is the sum of all the average powers over the frequency bandwidth. The power measurement of signals with no visible or prominent carriers requires the use of averaging techniques over time and frequency. Power bandwidth is the total average power of a signal contained in a specified frequency range.

Maintaining the proper average transmission power is a key adjustment made in cable systems for digital as well as analog signals. The most practical equipment for average power measurement includes SLMs, spectrum analyzers, cable TV analyzers, and power meters. A number of procedures using the built-in measurement tools of spectrum analyzers make average power measurements routine and consistent. The power meter requires a bandpass filter or an isolated channel for accurate channel measurements, but it provides the best accuracy of all the equipment discussed.

The computation of system total power for analog systems is well understood. Analog signal contributions are estimated from the peak carrier powers, and digital signal powers are as measured with average power techniques. The total system power is arrived at by adding the Watts for each signal.

Questions for review

1. The RMS voltage of an AC signal (True or false for each item)
   a) causes the same resistor heating as a DC voltage of the same value
   b) is always greater than the peak voltage
   c) requires a periodic voltage waveform and a constant load to calculate
   d) is dependent upon the shape of the signal
   e) is always the same as average power.

2. An analog amplitude modulated signal power changes as the information transmitted changes. Digital signal power remains constant over time; the information transmitted does not change the power added by the modulation. True or False?

3. The power bandwidth of a digital signal (choose all that are true)
   a) determines whether the signal is staying within its channel
   b) is used most often in the measurement of wireless communications
   c) is usually taken for granted in cable digital signals because the signal modulators must conform to strict bandwidth standards
   d) must be measured to make accurate measurements with a spectrum analyzer

4. The most accurate way to measure digital channel power is
   a) with a power meter tuned to the signal center frequency
   b) with a power meter whose input is restricted to the channel

bandwidth by a bandpass filter

c) power meters are not suited to measuring digital signal power

5. Measuring digital channel power with a spectrum analyzer can be done with (true or false for each one)

a) the analyzer's built-in channel power function provided you adjust the analyzer's resolution and video bandwidths for the specific type of digital signal, and a correction factor.

b) the analyzer's noise marker, provided you adjust the analyzer's resolution and video bandwidths to conform to the signal's transmission bandwidth, and apply a correction factor to account for the transmission bandwidth shape.

c) the analyzer as a manually tuned receiver to collect the channel power information across its bandwidth

6. To measure digital channel power with a signal level meter, the SLM must have preset filters and algorithms built-in for each type of digital signal to measure. True or False?

Selected bibliography

1. Helen Chen, "Testing Digital Video: New Measures of Signal Quality," *Communications Technology* magazine, Phillips Business Information Inc., Rockville MD, June 1995.

2. Dana Cervenka, "Designers Pour Smarts into Digital Test Gear," *Communications Engineering & Design*, Chilton Publications, New York, October 1996.

3. Jack Moran, personal communication.

4. M. Stephen McConnell, "Effects of analog and digital signals," *Communications Engineering & Design*, Chilton Publications, New York, December 1996.

5. Donald Raskin and Dean Stoneback, *Broadband Return System for Hybrid Fiber/Coax Cable TV Networks*, ISBN 0-13-636515-9, Prentice Hall PTR, Upper Saddle River NJ, 1998.

6. Boyd Shaw, *Power Measurement Basics*, 1997 Back to Basics Seminar, publication number 5965-7919E, Hewlett-Packard Company, April 1997.

7. Oleh Sniczko, "HFC Level Measurement Procedures," TCI Communications, Inc., sent to Bill Morgan of Hewlett-Packard Company, February 27, 1997.

8. Jeffrey L. Thomas, *Cable Television Proof-of-Performance; A Practical Guide to Cable TV Compliance Measurements Using a Spectrum Analyzer*, ISBN 3-13-306382-8, Hewlett-Packard Press, Prentice Hall PTR, Upper Saddle River NJ, 1995.

9. Robert A. Witte, *Electronic Test Instruments*, ISBN 0-13-253147-X, Hewlett-Packard Professional Books, Prentice-Hall, Inc., Englewood Cliffs, New Jersey, 1993.

10. Clyde F. Coombs Jr. *Electronic Instrument Handbook*, 2nd Edition, ISBN 0-07-012616-X. McGraw-Hill, Inc., New York, 1995.

10

Peak, Peak-to-Average, and Burst Power Measurements

Power that peaks or starts and stops randomly causes two problems. First, the average power, critical for system design and maintenance, is difficult to measure. Second, extreme power peaks cause signals to interfere with one another. This chapter helps you understand the nature of time-variable power by explaining why peaks occur in digital signals, and how to measure the peaks and average powers of digital and burst digital signals.

What you will learn

- Why does signal power variation affect system performance?
- What causes the power peaks in a digital signal?
- When is it important to measure peak power?
- How to measure peak power and burst power

The influence of peak and burst power on signal quality

BER increases when peaks are clipped, signals are interfered with, or the transmission path becomes overloaded. Figure 10.1 shows the influences of the peak and burst power signals on the digital signal quality. Burst signals influence total system power dynamically. The system design must be robust enough to prevent this bursting from causing the transmission system from overloading due to compression. Burst signals are also notorious for creating out-of-channel distortion. The peaking of power in a digital signal can cause interference with neighboring analog and digital channels if their relative power levels are set improperly. It is important to set the analog to digital signal relative powers to design standards, and keep them there. In the return path, setting RF or laser levels properly is an important factor in keeping BER low.

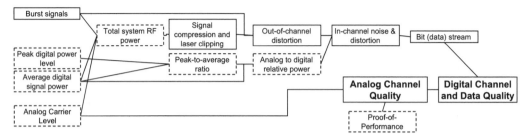

Figure 10.1. Effects of peak power on signal quality.

The nature of non-continuous power

Strong, non-continuous signals can imprint their patterns on other signals in the system. In contrast, when the power of an analog signal changes over time, the changes usually do not affect system performance. The system components are designed to take such changes in stride. However, if the change in signal power is too radical, for instance, if many of the analog signals carry exactly the same signal, synchronized with the same frame timing, the distribution equipment responds by imprinting all signals in the band with the power changes of the synchronized channels. This is the fundamental mechanism of cross modulation.

Non-continuous power changes over time. If the change is so radical it dips close to zero Watts, then the signal can be called a burst signal. Burst signals are usually used for return path communications. If the power of a signal changes quickly above the average power, it is called a peak. The peak responses of a digital signal are defined by the signal's peak-to-average power ratio. Peak-to-average ratio is discussed later in this chapter.

Defining peak, pulse, burst, and impulse

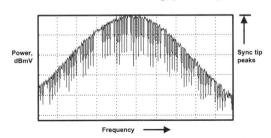

Power, dBmV

Sync tip peaks

Frequency ⟶

Peak, pulse, burst, and impulse are terms for the time domain behavior of a signal. Peak means the highest level attained over a given time period. Another term used is peak envelope power, because the peaks represent the extreme swings of energy whether they last long enough for one or more cycles of the carrier vector rotation or not. The traditional measure of an analog video carrier level is the signal's peak response, as shown in the figure at the left. The picture is from a spectrum analyzer zoomed into the visual carrier with a sufficiently wide resolution bandwidth (RBW) and fast enough sweep to see the time variations of the signal. These peaks are the signal's horizontal sync pulses. To estimate the RBW, knowing the signal's burst time, pick the bandwidth wider than 1.5/(burst time). For example, the NTSC video signal pulses are 4.7 µsec. The value 1.5/4.7µsec = 319.2 kHz. The analyzer's next highest RBW is 1 MHz.

In a digital signal, the peak values are predictable only on a statistical basis. From microsecond to microsecond peaks may or may not occur; they are not periodic. To measure the peak envelope of the analog video, the receiver has to have an IF bandwidth wide enough to allow the voltage swings to reach their peaks. This is no problem when the peak level is sustained for nearly 5 µseconds, but in digital signals, the peaks are more random and shorter in duration. Spikes of energy occur when the voltage waveforms of multiple signals happen to line up in time. The receiver bandwidth must be very wide.

Pulse is a term usually reserved for the controlled-time envelope of a carrier, such as the way a television sync pulse carves out the video carrier signal as it traces out the pulse's front porch, maximum carrier level, and back porch. The sync pulse maximums of an analog visual carrier are also called the peak responses of that signal for the purpose of measuring carrier levels.

Burst signals are defined here as a controlled-envelope signal which has information modulated onto it, and turns on and off according to traffic requirements.

Impulse power is a term usually reserved for interference that is the unwanted signal generated by an electrical or electro-mechanical process. Examples include automobile ignitions, refrigerator motors, and electrical switches. Some ingress is considered impulse, but the distortion products from burst signals can cause impulse-like interference. When measuring distortion, it is difficult to distinguish between burst, noise, and impulse.

Peak power and peak-to-average power ratio

The peaks in a digitally modulated signal come from the changes in the signal's amplitude as it moves from symbol point to symbol point. These peaks are a necessary part of digital modulation. Here is why:

- Digital signals are filtered to keep them within an allocated channel bandwidth. Certain symbol sequences cause occasional high-voltage peaks, but still stay within the channel bandwidth.
- Every specified modulation format configuration specifies the

signal's maximum peak-to-average power ratio, in dB.

- Increasing the average power of the digital signal increases the signal's peaks, dB for dB. These peaks interfere with analog signals and cause laser amplifier clipping.
- Setting the average power levels too high causes interference with the analog signals and laser amplifier clipping.

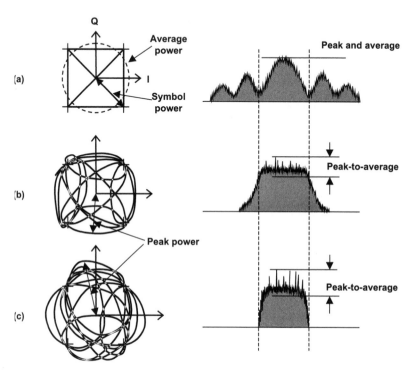

Figure 10.2. The effect of filtering on a QPSK digital signal.

Filtering narrows transmission bandwidth, increases peaks

Filtering affects a digital signal by smoothing the transitions between symbols, as shown in Figure 10.2. In (a), the digital information is not filtered, so the transitions between the symbol points are close to being straight lines. As the vector moves between points it makes sharp turns allowing it to reach the next symbol point. These transitions take a wide bandwidth to transmit because of the quick phase and magnitude changes in direction by the vector. Remember how many harmonics it takes to reconstruct a square-wave waveform? The signal's bandwidth, shown on the right, is very wide to accommodate the quick changes. The signal's average power is about the same or less than the symbol amplitudes. A filter is applied that smooths the sharp transitions, as in (b). The transmission bandwidth narrows because the sharp corners are removed, but the vector must be allowed to loop out from the symbol radius, causing peak amplitude responses above the average symbol power. The transitions must be made in the same amount of time, no matter what path they take. If the path is not direct, the longer path requires more voltage, and therefore power, to make up for the indirect path. Further filtering, as in (c), allows the signal to have even lower bandwidth, but the signal now has peaks further from the average symbol power.

Filtering a digital signal	Filtering a digital signal narrows its bandwidth, which makes better use of the frequency bandwidth. The side effects include increasing the peak-to-average power ratio and improving immunity to interference.

The lower bandwidth in-channel means that the signal trajectories, that is, the movements are from one symbol to the next, is less direct. For the same symbol clock accuracy, a lazy trajectory may miss the symbol mark more often than before, causing mistakes in reading the symbol. This effect causes intersymbol interference, or ISI. Modulation formats with higher numbers of symbols packed in the same channel envelope are closer together in terms of their I and Q spacing. This can cause an increase in sensitivity to ISI, and therefore BER. Formats with more than 64 symbols are especially susceptible to increased error unless they are transmitted in paths free from impairments, noise, and interference.

Intersymbol interference (ISI)

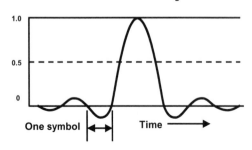

Symbols are prevented from overlapping one another by filtering them with a filter designed to oscillate, or ring, at the symbol rate. Here is how it works. The drawing shows the response of the filter in time. Its maximum is equal to the amplitude of the symbol, and its pre- and post-responses, the waves to either side of the main response, go through zero at exactly the timing of the symbol before and the symbol following the one currently in the filter. When the symbol is transmitted, the effects of the adjacent symbols, in time, are being cancelled by the filter because the voltage is zero at the symbol times. If the filter is not exact in its transitions through zero to match the symbol time, then the voltage nudges the symbol position, and creates ISI. Filters that have the ability to cross zero voltage at the symbol points are Nyquist, raised cosine, square-root raised cosine, and Gaussian.

Usually the filter is split, half being in the transmit path and half in the receiver path. This allows a fast transition in a signal to occur within the occupied bandwidth, reduces the tendency of one signal or one transmitter to interfere with another in a frequency division multiple access system, such as a cable system. On the receiver end, the filter provides reduced bandwidth, which improves sensitivity with the rejection of noise and interference.

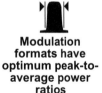

Modulation formats have optimum peak-to-average power ratios

Each digital format has a range of peak-to-RMS or average ratios that optimize the transmission quality of that specific modulation format. The peak power can be as much as 6 to 15 dB higher than the average power, depending upon the modulation format. Table 10.1 shows some examples. The variations are meant to illustrate the wide range of the way the format can be used, not to imply that these variations take place in any one application. In fact, the peak-to-average ratio of digital modulators is a rigorously maintained specification. Measuring this ratio is the subject of procedures later in this chapter.

The peak of a CW signal

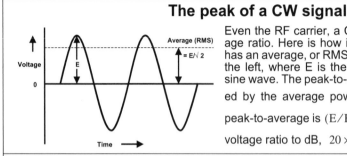

Even the RF carrier, a CW signal, has a peak-to-average ratio. Here is how it is calculated. The CW signal has an average, or RMS, power shown in the picture on the left, where E is the maximum swing of the signal sine wave. The peak-to-average voltage ratio is E divided by the average power, which is $E/(\sqrt{2})$. So the peak-to-average is $(E/E/(\sqrt{2})) = \sqrt{2}$. Converting this voltage ratio to dB, $20 \times \log \sqrt{2}$, or 3 dB.

Different symbol patterns, symbol clock speeds, channel filter, and channel bandwidths are conditions which set the peak-to-average ratio. These peaks can cause amplifier compression, resulting in intermodulation interference in the cable system. Excessive peak powers also cause optical fiber system amplifiers to overload, causing the signal to be clipped of these valuable peaks. The effect on modulation quality is similar to that of amplifier compression as shown in Figure 8.6 on page 136.

Format	Peak-to-average ratio
CW signal	3 dB
QPSK	3 to 6 dB
64-QAM	5 to 8 dB
VSB	8 dB
OFDM	7 dB
CDMA	9 to 15 dB
Unfiltered noise to 1 GHz	8 dB

Table 10.1. Typical peak-to-average ratios for various formats.

Power levels for digital signals are set using their average power. If the peak-to-average power ratio is higher than the design parameters allow, the result is distortion and interference. For example, the 64-QAM signal used in DVB-C systems is defined and expected to lie within the 5 dB to 8 dB range, if the peaks go to 9 or 10 dB, system BER increases.

If the peak-to-average increases on the incoming signal, there is little you can do but complain to the service providing the signal. If you suspect that the signal is being distorted by your distribution system, look for excessive adaptive equalization at the receivers. This indicates frequency response and/or group delay problems as discussed in Digital Signal Quality, Chapter 8 beginning on page 127.

Measuring peak-to-average ratio

Digital signal peaks are fast and random

The peak power of a digital video signal is random, fast, and occasionally very high. The random nature of the encoded digital modulation insures that the average power of the RF signal is consistent. The peak power of the signal's RF envelope, unlike an analog video signal, has to be viewed as a statistical measurement. Once in a long while a particular data stream sequence results in a high peak. Peak power measurements taken over identical time spans may give different readings. In practice, however, peak power readings reach a reasonably consistent level within seconds.

Specialized baseband test equipment is required to make peak-to-average ratio accurately. The peak-to-average is simply a comparison of the separate absolute peak and average power readings. Measuring the average power is covered in Average Power Measurements, Chapter 9 beginning on page 155. Peak power is measured with a **peak power meter**, a **digital video power analyzer**, or a **spectrum analyzer**. Dedicated digital modulation and video power analyzers provide direct ratio measurements as well as statistical information about the frequency of the peak signal occurrences. Peak power meters and video power analyzers provide the receiver bandwidths necessary to capture very fast peaks; peaks that occur in 200 ns, that is, 200×10^{-9} seconds duration.

Spectrum analyzers do not make peak measurements

The spectrum analyzer is not equipped with an IF bandwidth shape and signal capturing detection necessary to catch these peaks directly, although it can calibrated to make useful measurements. An excellent example is Figure 10.3. This graph shows the peak detected response of a noise signal as a function of the observation time, product of τ and RBW, where τ is either the length of the RF burst or the spectrum analyzer sweep time divided by the number of cells in the sweep. As the burst lingers longer in the analyzer's IF, either because the RBW is wide or the sweep is slower, the detected peak increases. For more information see Gorin in this chapter's bibliography.

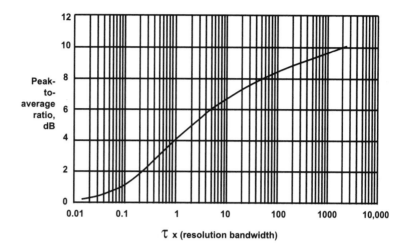

Figure 10.3. Spectrum analyzer level setting for peak-to-average measurements.

Here are a number of procedures to help you understand the application of a spectrum analyzer to the measurement of burst signals. The following illustrates some of the conditions that need satisfying to use a spectrum analyzer. Don't be intimidated by the list of conditions in the next procedure. The procedures provide specific rules to follow.

Tips on measuring peak digital signal power with a spectrum analyzer

1. The spectrum analyzer's sweep lowers its chances of it being tuned to the frequency of the peak when it occurs. Digital signal peaks occur as the result of specific code sequences, so they do not happen in a predictable period. These facts make it necessary to gather data for a long period of time.
2. The digital signal must be at least 10 dB out of the displayed average noise level to prevent measurement errors. Low-level signals may require boosting with a high-quality preamplifier.
3. Measuring the signal's average power near the top of the analyzer's display risks having the analyzer clip the peaks of the signal. To avoid this, use the reference level to bring the signal power down at least 10 dB from the top reference level.
4. If the analyzer is overloaded, peaks can be compressed, that is, measured significantly lower than actual. Use a bandpass filter to reduce the total RF power to the analyzer's input mixer.

5. The IF bandwidth filter needs to be set wider than bandwidth predicted by the fastest peak response. For example, if the peak response is known to be 20 µs, the bandwidth would have to be set to more 1.5/(20 µs) using the rule in "Defining peak, pulse, burst, and impulse" on page 193. The RBW needs to be >75 kHz. For a 5 MHz symbol rate the burst is 200 ns long. The RBW needs to be 1.5/(200 ns), or 7.5 MHz. Few spectrum analyzers offer such a wide bandwidth. You would need to use a vector signal analyzer. A vector signal analyzer offers a wide receiver bandwidth and very fast digital signal processing. See "Vector signal analyzer" on page 260 for a brief discussion of its capabilities.

6. The video bandwidth has a great influence on burst measurement results because of its influence on the RBW response. The following procedures help establish some working rules for its use.

7. Peak data gathered with the spectrum analyzer may not provide accurate absolute power, but its results can be used to compare the relative peak responses of two like-formatted signals.

Measuring burst power

Radical changes in signal power can cause amplifier compression, cross modulation, and transient distortion. A signal intentionally turned on and off, such as from a cable modem, is a **burst** signal. Chapter 11 beginning on page 211 covers the interference aspects of intermittent signals.

Burst signals are used to send information efficiently

Burst power is the use of non-continuous power to send information, but only when the signal is on. During this on period the signal operates like any other digital signal, although usually in a narrower bandwidth. The off periods are not just a cessation of data or modulation, but a total shut off of the signal power transmission. Today's main use of burst signals is for data transmission in the return path of the cable, commonly from 5 to 42 MHz band in most North American systems. A number of strategies are being integrated into the cable system to make the most of the return path for digital communications.

Measuring the average power of burst signals is difficult

All RF systems have to work within certain power transmission boundaries to meet their designed performance. Setting the average power of the continuous analog and digital signals assures that every system section operates at its designed performance. But what if these signals were changing constantly? Radically changing levels in the forward path would cause chaos with all the feedback mechanisms designed to stabilize the gain, flatness, and therefore, the distortion performance at the subscribers' receiver.

Burst power is tough to measure because the measurement instrument has to capture the signal, store the results in memory, and accumulate results over time. The most convenient and available tool for measuring burst power is the signal level meter. It has sufficient bandwidth, data collection techniques, and is easy to use. Part of its ease is that there are not RBW or VBW adjustments to make; the meter has already been calibrated for its intended digital signal.

The generic spectrum analyzer does not have the ability to record a long period of signal waveform in a digital memory for later analysis, as do vector signal analyzers. But the spectrum analyzer is convenient and available to most systems. It provides good measurements as long as the measurement procedures are kept consistent. The following procedures make burst measurements using a Hewlett-Packard Cable television personality (the personality card, model HP 85721). The measurements include the average burst power of a single channel, the average burst power of an entire return path band, the burst power of a signal channel using the analyzer's zero span, and then confirmation of the signal's peak-to-average correction factor.

Here is how to measure a single burst channel with the cable television personality channel power function.

Measuring channel burst power HP 85721 cable TV personality channel power function

1. Set the span to see a bit more than one channel. For example, if the burst signal is a 1280 ksymbol/second QPSK signal, in a 1.28 MHz channel bandwidth, set the span to 2.0 MHz.
2. Use the digital channel pwer key to start the function.
3. The routine sets the analyzer for RBW of 30 kHz, VBW of 300 kHz, and sample detection. Sample detection assures that the analyzer records the amplitude value at the end of each frequency slot, not the peak value, as it does in the normal CW measurement mode.
4. The program automatically integrates the power across the channel and presents a results of a continuing measurement.

5. Save and accumulate amplitude information by setting the trace to maximum hold. As the bursts are caught over the entire channel, the trace fills in completely. The measurement is complete when the amplitude buildup slows, usually in a few seconds, as shown in Figure 10.4

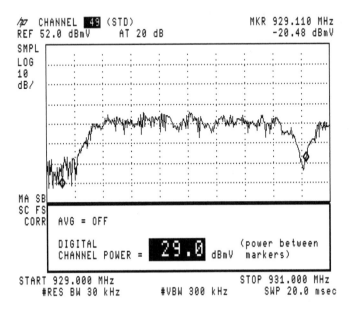

Figure 10.4. QPSK burst signal power.

6. The amplitude is adjusted for the peak-to-average ratio by subtracting 4 dB from the accumulated peak power of the channel. The power is 29.0 dBmV - 4 dB = 25 dBmV. The technique for measuring the value for the peak-to-average correction is measured later in this chapter. Note that this measurement looks at the highest TDMA transmitters in the channel. Later in this chapter is a procedure for examining individual transmitters, the zero span method.

This measurement can be made using the analyzer's built-in channel power function, found under the MEAS/USER menu in Hewlett-Packard 8590E-Series analyzers, although care must be taken to make the same setting decisions as does the cable TV software.

Here is how to measure the total average power of a number of burst signals in the return path.

Measuring return path power with spectrum analyzer HP 85721 cable TV personality channel power function

1. Determine the full return path peak power with burst carriers and correct the value for average power. For example, use 20 burst signals as in the last two procedures, spread over 35 MHz of the return path bandwidth. (For this measurement, the signal source was a 20 channel synthesized signal from an arbitrary waveform generator set to a 900 MHz center frequency.)

2. If all the carriers are the same amplitude, the total power is the average power plus 10 log 20 channels = 13 dB.

3. Set the span to see all the return bandwidth. Use 35 MHz as an example.

4. Set the center frequency for the middle of the return path. For a 5 to 40 MHz return path the center frequency is ((40-5)/2) = 17.5 MHz.

5. Select the channel power measurement.

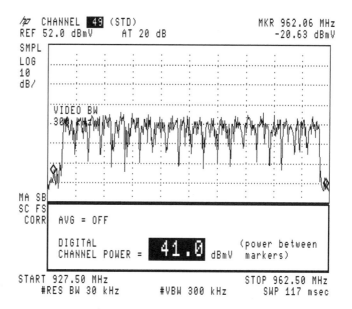

Figure 10.5. Average power for 20 burst signals in the return path.

6. The routine sets the analyzer for RBW of 300 kHz, VBW of 3 MHz, and sample detection. Sample detection assures that the analyzer records the amplitude value at the end of each frequency slot, not the peak value, as it does in the normal CW measurement mode. To use this routine as if testing a single channel, use the bandwidth key to reset the RBW to 30 kHz and the VBW to 300 kHz.
7. The program automatically integrates the power across the channel and presents the results of a continuing measurement.
8. The continuous measurement means that when the burst is missing, so is its effect in the results. To save and accumulate amplitude information, set the trace to maximum hold. As the bursts are caught over the entire band, the trace fills in to show the individual burst signals. This may take only 10 sweeps. New bursts change amplitude by only 0.1 dB.
9. Read the peak value as 41 dBmV, as shown in Figure 10.5
10. The amplitude is adjusted for the peak-to-average ratio to get the average power by subtracting 4 dB from the accumulated peak power of the channel. The average power is 41 dBmV - 4 dB, or 37 dBmV.
11. Since the individual signals in this example are from the previous procedure, the total power of 20 like signals is 10 log (20), or 13 dB, higher. Add 13 dB to previous procedure's results. 25 dBmV + 13 dB is 38 dBmV, close to the 37 dBmV results in this test.

The next procedure measures the power of the single burst signal and confirms the 4 dB peak-to-average ratio correction using the analyzer in its zero span to use the analyzer in its tuned receiver, mode.

Measuring burst power of a single channel with a spectrum analyzer in zero span using the noise marker

1. Measure a single QPSK burst carrier with a 3-dB transmission bandwidth of 1.28 MHz and a channel spacing of 2 MHz.
2. Center Frequency is set to center of the desired channel.
3. Set the analyzer to zero span.
4. The RBW is set using the equation found in "Setting spectrum analyzer bandwidths for digital power measurements" on page 179. The RBW = 30 kHz.
5. Set the reference level so the response does not go above the top line or close to the bottom graticule.

6. The video bandwidth must be sufficient to smooth the burst without allowing wide voltage excursions. The video bandwidth affects and is affected by the RBW. A good rule is to use VBW = 0.35/(burst on time) = 0.35/250 μsec = 1.4 kHz. Use the next largest bandwidth, 3 kHz.

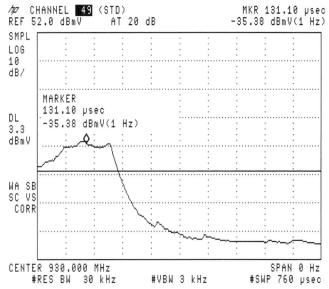

7. Set the sweep time to greater than 3 times the burst time, or 3 x 250 μsec = 750 μsec.
8. Set to video and adjust the trigger level to get a response.
9. Select the single sweep mode to capture one pulse.
10. Use noise marker on the pulse and record the power level, -35.38 dBmV.
11. Correct the power for the channel's 3 dB bandwidth. For a 1.28 MHz burst channel bandwidth, and a spectrum analyzer noise marker referred to 1 Hz, the correction is 10 log (1.28 MHz/1 Hz) = +61.1 dB. The power is 61.1 - 35.38 dBmV = 25.7 dBmV
12. Now confirm the peak-to-average used in the preceding procedures. Set up the same signal with a 30 kHz RBW and a 300 kHz VBW.
13. Set the sweep time to 300 μs and trigger a single sweep to capture a single burst.

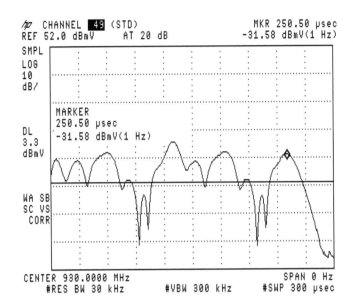

Figure 10.6. Confirming the peak-to-average ratio of a burst signal.

14. Place the noise marker on one of the highest peak responses, shown in Figure 10.6, and compare its value to that found in Figure 10.5. The -35.38 - (31.58) is about 4 dB.

15. Note: the noise marker averages 16 points on either side, , or about 8/10s of a division. The amplitude averaging makes the marker appear to be lower than actual.

For TDMA signals the above procedure can help you see both transmitter power and ingress/system noise. To see specific transmitters, set the sweep time to the frame time of the individual TDMA transmitters. In between these transmissions the power measurement is made on either ingress in the channel, or the system noise floor. Since the noise marker averages over 0.8 of a division use the normal marker for these additional measurements.

Peak and burst measurements

Peak and burst measurements are required to keep track of the average power in a system and to prevent exceeding design limits, especially for laser hubs.

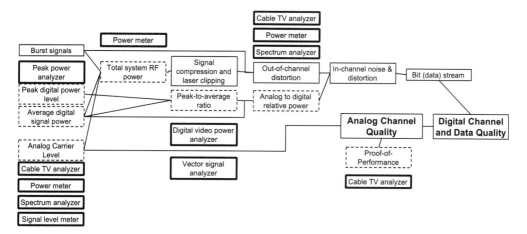

Figure 10.7. The test equipment for measuring peak and burst powers.

Each piece of test equipment in Figure 10.7 requires specific procedures and/or test conditions to make accurate and repeatable measurements.

Summary

Non-continuous power causes some unique system and measurement problems. BER increases when peaks are clipped, signals are interfered with, or the transmission path becomes overloaded. Strong, non-continuous signals can imprint their patterns on other signals in the system. Burst signals are usually used for return path communications. If the power of a signal changes quickly above the average power, it is called a peak.

The peaks in a digitally modulated signal come from the changes in the signal's amplitude as it moves from symbol point to symbol point. These peaks are a necessary part of digital modulation. Filtering affects a digital signal by smoothing the transitions between symbols. Each digital format has a range of peak-to-average ratios that optimize the transmission quality of that specific modulation format. Power levels for digital signals are set using their average power.

The peak power of a digital video signal is random, fast, and occasionally very high. Specialized baseband test equipment is required to measure peak-to-average ratio accurately. The spectrum analyzer is not equipped with an IF bandwidth and signal capturing detection necessary to catch these peaks directly, although it can make approximations useful for making comparative measurements.

Radical changes in signal power can cause amplifier compression, cross modulation, and transient distortion. Burst power is the use of non-continuous power to send information. Burst power is tough to measure because the measurement instrument has to capture the signal, store the results in memory, and accumulate results over time. The generic spectrum analyzer does not have the ability to record a long period of signal waveform in a digital memory for later analysis, as do vector signal analyzers.

Questions for review

1. Choose all that apply to the definition of peak power.
   a) the maximum power spikes in a digital signal
   b) the maximum envelope of the signal's amplitude
   c) the carrier level in an analog visual carrier
   d) all of the above

2. Peak power is measured by (choose all that apply)
   a) Peak power meter
   b) Modulation domain analyzer
   c) Oscilloscope
   d) General purpose spectrum analyzer
   e) Digital video signal analyzer
   f) None of the above

3. A 12 dB peak-to-average ratio is considered high. In a digital signal this high ratio suggests
   a) an efficiently operating digital modulator
   b) good modulation quality
   c) the peak power is 12 dB above the average power

4. In setting the relative levels between analog video and digital video signals in a broadband cable system, use the peak power of the digital signal. True or False.

5. Burst power is used to send digital information in the return path in a cable television system. Why are burst signals used? Choose all that apply.
   a) Burst signals are less susceptible to interference.
   b) Burst signals are used in the time and frequency domain multiplexing, required to optimize the efficient use of the return path bandwidth.
   c) They are the only way narrow channel bandwidths can be used.

Selected Bibliography

1. Helen Chen, "Testing Digital Video: New Measures of Signal Quality," *Communications Technology* magazine, Phillips Business Information Inc., Rockville MD, June 1995.

2. Roger L. Freeman, *Telecommunication Transmission Handbook*, 3rd edition, ISBN 0-471-51816-6, John Wiley and Sons, Inc.,1991.

3. Joe Gorin, *Spectrum Analyzer Measurements and Noise*, Application Note 1303, Hewlett-Packard Company, publication number 5966-4008E, May 1998.

4. Ron Hranac, "Making Two-Way Work (Part II)," Society of Cable Telecommunication Engineers, Cable-Tech Expo '97, *Proceedings Manual: Collected Technical Papers*, Volume Two, 1997.

5. Donald Raskin and Dean Stoneback, *Broadband Return System for Hybrid Fiber/Coax Cable TV Networks*, ISBN 0-13-636515-9, Prentice Hall PTR, Upper Saddle River NJ, 1998.

6. Deiter Scherer, "Measurement Tools for Digital Video Transmission," Hewlett-Packard Video Communications Division, *IEEE Transactions on Broadcasting*, Volume 39, No. 4, December 1993.

11

Distortion, Noise, and Interference

Distortion, **noise**, and **interference** are closely related. Digital signals produce distortion that is noise-like because the digital signals themselves are like noise. Distortion and ingress interfere with the signal in-channel as does system noise. It is difficult to determine whether signal quality problems come from distortion, noise, or interference. But it is important to be able to trace and correct their sources. This chapter discusses the additional effects that these culprits have over and above acceptable analog system performance.

What you will learn

- The differences between noise and distortion sources in systems with digital signals.
- What are the effects of distortion, noise, and interference on digital signals?
- What instruments help measure and troubleshoot?

Effects on digital signals

Distortion, noise, and interference cause crashes

Distortion, noise, and interference affect digital signals by eroding the digital information encoded inside them until the signal can no longer sustain its high quality, and it goes over the cliff. Analog signals display a slow decay in quality as conditions degrade the signal content. The signals that cause this degradation in analog and digital signals are distortion and noise products of other signals in the system, transport conditions, and unwanted signals from outside the system.

Figure 11.1 shows a map of the system signal performance parameters that influence distortion, noise, and interference. Starting on the left, high signal powers cause signal compression. The peaks of digital signals are flattened, increasing BER as the symbol points are missed by wayward vector trajectories. Compression also causes distortion products outside the channel which is compressed. CSO and CTB are the distortions created by analog signals in this manner. Burst signals, discussed in the last chapter, give rise to distortion, usually outside the burst channel itself. Digital signal distortions, caused by the mixing of digital signals are noise-like, and are referred to as composite intermodulation noise, or CIN. Noise and interference are next, adding to the in-channel distortion and noise as an "ambient" background affecting MER. Group delay is a prominent transport impairment which distorts the digital signal in its own channel.

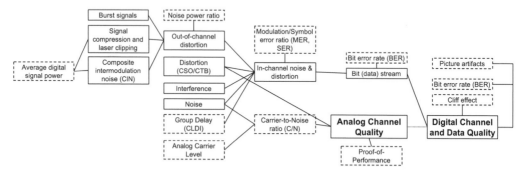

Figure 11.1. Map of the influences of distortion, noise, and interference on digital signals.

The lower part of the figure shows the analog quality path where C/N and the entire host of analog North American proof-of-performance standards, sometimes abbreviated POP, help determine the signal quality along the way. Just as in signal power and system flatness, distortion, noise, and interference, the starting point for good digital signal quality is a necessary foundation for analog signal performance. Without analog signal quality, it is nearly impossible to troubleshoot and diagnose digital signal quality problems.

Digital signal quality starts with POP	As a minimum requirement for good digital signal quality, your system must be up to the good analog measures, called proof-of-performance, or POP, in North America. Without this fundamental quality in carrier-to-noise, CTB, and CSO, maintaining, troubleshooting, and diagnosing digital quality problems are impossible.

Intermodulation distortion

Distortion from digital signals looks like system noise

Intermodulation distortion creates unwanted signals from the mixing of desired and undesired signals. These distortion products interfere with the other transported signals. The unwanted signals are outside the channels that create them, that is, out-of-channel distortion, just like the familiar CSO and CTB named for the second- and third- order distortions created by analog signals. In systems where every channel is occupied, the distortion products fall in the channel of another signal, disrupting its quality.

The distortion caused by digital channels is much more noise-like than the composite distortion of analog signals, CSO and CTB, because they are generated by mixing digital signals together, as illustrated in Figure 11.2. This digital distortion has been given a new name, **composite intermodulation noise**, or **CIN**. As the name implies, the energy is a product of mixing signals, but it looks like noise. Composite means the same as it does in CSO and CTB: the effect of several distortion products in the same bandwidth.

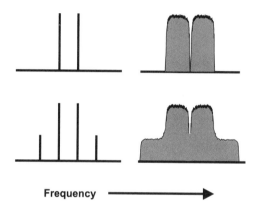

Figure 11.2. Distortion products of CW and digital signals.

In the over-simplified pictures in Figure 11.2 the top row is a pair of CW signals on the left and digital signals on the right with no visible distortion. The second row shows the effects of severe third-order intermodulation distortion applied to each signal pair. The CW signals generate close-in CW distortion products which are spaced equally on each side of the pair. The digital signals produce noise-like sidebands, centered on the third-order spacing, but overlapping the signal edges so that they give the original signals new sidebands which look very much like the parents. In a system with dozens of digital and analog signals the distortion would retain its noise-like appearance, but the mixing and re-mixing of legitimate and distortion signals, that is, the composite nature of the distortion products are predictably placed. (Noise sidebands like these would only appear in severely impaired cable systems. Such signals would have gone over the cliff long before these distortion levels were reached.)

CIN is not as predictable as CSO/CTB

The out-of-channel distortion products not only are noise-like, but they also appear at the frequencies throughout the channels in a random fashion, rather than in the more statistically predictable places CSO and CTB fall. This is because the uniform channel power of a digital signal mixes over wide bandwidths; in analog signals, the distortion produced by analog signals is primarily from mixing the strong video carriers of the PAL and NTSC signals.

CIN in the system

Such distortion products are called composite intermodulation noise, or CIN. This distortion/noise energy is the digital signal system equivalent of the analog CSO and CTB products.

Digital signals are susceptible to distortion over their whole bandwidth

Digital signals are susceptible to distortion anywhere in their channel. PAL or NTSC signal distortion susceptibility depends upon where in the channel band the distortion falls. But the digital signal has forward error correction and adaptive equalization to compensate for the non-linear CIN distortion just as they do for linear distortion. The FEC is adept at handling these noise-like distortions because the CIN is Gaussian in nature, while the adaptive equalizer is better suited to handle the analog distortions which appear as a cluster of beats.

Checking distortion at the modulators

Since the spectrum in most systems is so crowded there are few places to check the out-of-channel distortion created by most signals. One check is to measure the adjacent channel power, ACP, at the output of the digital modulators themselves using the spectrum analyzer ACP function. Many modulators process two channels at the same time, for cost-savings reasons, so a slight disadvantage is that only one side of each channel can be measured.

Adjacent channel leakage, ACL measures digital modulator distortion, or leakage in adjacent channels Small ratio indicates faults in the modulator's output filter or an overdriven condition. This is an in-service test if made from a modulator test port or directional coupler.

The measurement situation is made worse by the facts that digital signal-to-noise ratio measurements

    1) require an adjacent empty channels, and
    2) in-channel effects cannot be measured with non-interfering techniques similar to those used in the C/N for analog signals today.

With few empty channels the only opportunity to measure distortion is by interrupting service. Fortunately, you can rely on the dedicated signal quality measurements, such as MER, discussed in Digital Signal Quality, Chapter 8 beginning on page 127.

No in-service tests for distortion on digital signals

The second point is a disappointment to those familiar with current proof-of-performance measurements for PAL and NTSC systems. In these measurements, the test equipment can synchronize with the modulation of the television signal under test, virtually eliminating the signal's modulation, as if the channel were either off-air, or unmodulated. These non-interfering, or in-service, tests of C/N and distortion are conducted without tampering with the transmission in any way except to make sure there is a blank interval line or two. Digital signals offer no such "look through" opportunities. Their signal information is buried beneath layers of signal processing such that subtle noise and distortion disturbances beneath the signal information cannot be extracted by the traditional analog techniques. These disturbances can be seen by constellation diagrams and the degradation of EVM and MER. With currently available test equipment the signal must be turned off in a channel to see distortion and noise beneath.

Compression

Compression of an active device, such as a laser or trunk amplifier, is the result of putting more power into the device than its design allows. The amplifier's gain is the ratio between the power in and power out. Ideally, the linear relationship between power in and power out is a straight line in this illustration. Linear means that for every change in input power there is a corresponding and predictable change in output power, hence the straight line on the power curve.

When the amplifier has more power input than its circuits can handle, the power out no longer matches the power input, and a roll-off, shown in the non-linear portion of the curve, is the result. The gain of the amplifier is still at work, but can no longer transfer this energy to the fundamental signal, as shown by the frequency domain sketches. Instead, the extra energy, shown as the cross-hatched area between the output curve and the dashed line, goes into harmonics and intermodulation distortion products. As the level of compression increases, these distortion signals rise, usually faster than the fundamental signal.

There is usually non-linearity at the low-power end of the curve also. When the input signal is very small, at turn-on, the amplifier is operating with gain changing as a function of the input level.

Systems that include many digital channels experience noise-like, that is, **non-coherent**, composite distortion products. If an adjacent channel is without a signal, the signal-to-distortion can be measured using the adjacent channel power functions discussed in Average Power Measurements, Chapter 9 beginning on page 155. Since the distortion from digital signals is like noise, then the signal-to-distortion ratio becomes a signal-to-distortion-plus-noise measurement.

Noise and non-coherent distortion have the same effect on channel performance.

In- and out-of-channel distortion

The out-of-channel distortion created by one channel is the in-channel distortion of other channels when the distortion product falls in the channel's bandwidth. A signal being compressed by an amplifier can also show distortion internally, as the energy of the highest-magnitude symbols is reduced to power the intermodulation and harmonic distortion products. The third-order distortion products can fall inside their own channel bandwidth, appearing like noise modulation on the symbols.

Use ACP to measure signal-to-distortion ratio

Adjacent channel power compares the average signal power of the active channel to the average power of either adjacent empty channel. The power can be measured as relative dB or as two absolute powers. Figure 11.3 shows the use of a spectrum analyzer to make an ACP measurement of a slightly compressed 64-QAM signal. Note the graphical display of spill-over in the adjacent channel. The signal-to-distortion-plus-noise is the center channel power of 63.2 dBmV. The display readout is configured to display the difference between the center channel and the adjacent channel powers, at -37.9 dB and -39.3 dB. The adjacent channels have 25.3 dBmV and 23.9 dBmV in them, respectively.

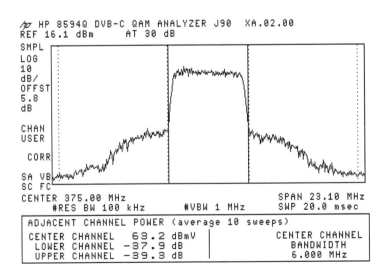

Figure 11.3. Absolute power measure of ACP on an isolated digital signal.

Distortion levels

Distortion tolerance: 30 to 40 dBc for non-coherent; 40 to 50 dBc for coherent levels

There are no standards for the signal-to-distortion-plus-noise ratio for digital signals. But experience has shown that levels between 30 and 40 dBc do not affect most digital signals up to and including the more complex 256-QAM formats. However, the less noise-like the distortion, that is, the more **coherent** the distortion, the more the signal is affected. The same distortion affects analog signals at much lower levels: a range between 40 and 50 dBc. The failure mode with the digital signal is a crash, whereas the analog signal slowly degrades as the distortion gets closer to the signal's level.

Coherent distortions include all the distortions associated with analog systems, that is, CTB, CSO, cross modulation, ingress interference, and head end-generated spurious. Coherent distortion within the channel bandwidth cause BER and SER increases; distortion close to the channel may or may not affect the digital reliability and margin depending upon the amount of signal data compression, its symbol rate, and its modulation complexity. Some NTSC analog modulators generate sufficient LO feedthrough, about 45 MHz offset from the modulated carrier, of sufficient strength to corrupt a digital signal. In addition, analog modulation also can produce an audio FM image 4.5 MHz below the visual carrier that can interfere with a lower adjacent digital signal. Usually these signals are not noticed in analog systems, or in systems using the highest quality analog modulators.

Keep digital signals 6 dB lower than analog carriers

For systems that have analog and digital signals grouped together by frequency, testing has shown that an analog to digital power ratio of 6 dB seems to allow the best performance of both the analog and the digital groups, although typically a difference of 10 dB is used. The digital signals, being non-coherent, have a strong influence on the system C/N and distortion levels for the above mentioned reasons. Bringing up the levels of the digital signals, that is, making them closer to the analog levels, dramatically increases the noise due to intermodulation. Analog signals do not seem to have as much affect on

the digital signals as digital signals have on analog signals. See Mc-
Connell in the bibliography at the end of this chapter.

Distortion and laser clipping

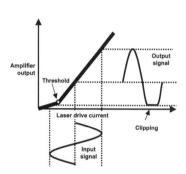

In HFC systems, the biggest contributor of distortion is laser overload due to high average power and power peaks
of the transmitted signals. This is especially true for the return path where signals are funneled together from the
multiple nodes that make up the optical system architecture. Just one out-of-specification laser can cause the all
the outputs of the node to be distorted.

Unlike conventional RF amplifiers, laser amplifiers have a
lower threshold which can distort the signal waveform if
the signal amplitude is too wide for the amplifier's dynamic
range. This is shown in the drawing at the left. As the input
signal grows in amplitude, the lower side is "clipped" off by
the non-linear portion of the power transfer curve. Each
cycle the laser has trouble turning on.

Overload due to the overdriving of the power to an amplifier, discussed in "Compression" on
page 215, causes lasers the same problem. In the case of compression, the top of the sine wave
is clipped, not the bottom. But the distortion effect is the same. Laser clipping causes high levels
of second-order distortion, but very little additional third-order distortion. Clipping can be a problem
in both forward path and return path transmissions, but because it is sensitive to signal power and
loading, the forward path flow, where the energy of signals from subscribers are additive, may be
the most troublesome areas.

Clipping causes a unique pattern to the observed constellation
patterns of QAM signals. Rather than a general smear of each
symbol, as shown in the noise cluster in the drawing on the right
of this paragraph, a minority of symbol dots seem to be shot out
of the symbol cluster, as shown in the clipping drawing. This
scattering is caused by the impulsive, or shot-like, nature of clipping.

Noise   **Clipping**

Much more information on the effects, and guidelines for avoid-
ing clipping is found in Raskin and Stoneback, a reference listed at the end of this chapter.

A new role for noise measurements

**Expect more
noise in systems
with digital
signals**

Carrier-to-noise ratio, traditionally held as the single most important
measure of system performance, is going to yield to MER as HFC
and cable systems are filled with digital signals. Although noise in a
cable system is trouble by itself, and it masks other system parameters critical to preventative maintenance, digital signals, because of
their protective layering, are less susceptible to noise than analog signals. For these signals, C/N is not a perfect measure of picture or data
quality for a given system. MER, as an in-channel signal-to-noise
measurement, and EVM, as a modulation quality measure, must be
used to judge the signal quality and its margin. See Digital Signal
Quality, Chapter 8 beginning on page 127.

C/N to monitor CIN levels

The measurement of noise level becomes more important as your system begins to fill up with digital signals because they add CIN. The system's analog signals C/N, requires more careful monitoring as CIN adds to the system amplifier noise. Noise affects analog signals directly and digital channel margin indirectly. In fact, monitoring analog signal C/N with a traditional tool, such as the cable TV analyzer, may allow you to keep track of the CIN distortion caused by the digital signals.

Noise rules are changing

The primary advantage of digital technology is its ability to overcome an analog transport system's gradual accumulation of noise and transmission impairments. Even the efficient transport of fiber optic cable contributes noise equivalent to about that of ten cascaded amplifiers. Once these cumulative impairments have been added to the desired video signal, they cannot be removed. Digital signals are susceptible to noise, too, but are more robust than analog signals when subjected to the same levels of noise.

Digital signals add a new layer of system noise

Noise contributions from digital signals can be thought of as adding a layer to a cake. The makeup of noise is shown in Figure 11.4. The cake plate upon which all these noise layers rest has no noise because it is frozen at zero degrees Kelvin. Zero noise power, -125 dBmV/Hz, is the reference for the real world. Above the reference is the noise contribution of the measurement equipment. When you make a measurement using a spectrum analyzer, the noise contributed by the analyzer is the line across the screen of the analyzer when you pull the cable off the front panel. It is the noise generated by the spectrum analyzer as a necessary part of its internal amplification and conversion processes. If the noise rises when you connect your cable system to the input of the analyzer tuned to an empty place in the cable spectrum, the noise displayed is the cable system noise. The next layer of noise is added by digital signal intermodulation of noise-like signals, composite intermodulation noise, or CIN. The next layer is the combined noise powers, upon which the signals, either analog or digital, reside.

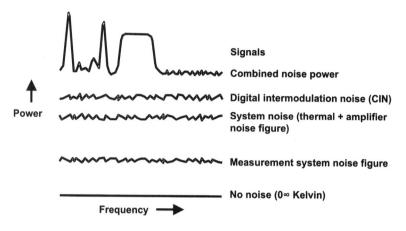

Figure 11.4. System noise in layers.

C/N is the difference between the analog carrier peak level and the combined noise power. MER looks at the effect of all these noise contributors on the quality of the symbol cluster. CIN is the ratio between the digital signal's average power and the combined noise power.

How noise affects bit rate

Noise sets a limit on bit rate. In 1948 C. E. Shannon, in an article for the Bell Systems Technical Journal, linked noise theory with digital transmissions. He said that the maximum data rate of a noisy channel whose bandwidth is H Hz and whose signal-to-noise ratio is S/N has a maximum bit rate of H x $\log_2$ (1+S/N) bits/second. For example, a telephone circuit with a bandwidth of 3925 Hz and a noise ratio of 39 dB can never transmit much more than 50,000 bits/second no matter how many or few signal levels are used and no matter how often or how infrequent the samples are taken. Shannon's equation gives an upper limit. Real systems rarely ever achieve it.

There are no noise level numbers in the figure because all system designs and measurement instrumentation configurations are different. These levels must be documented during design, installation, and maintenance of your system in order for you to provide consistently high quality for analog and digital signals alike. It is your job to maintain the power levels according to design specification for intermodulation distortion, especially the frequency response and input/output levels of amplifiers and fiber link interfaces. Use full system sweep techniques between head end and major hubs, not just the monitoring of a few channels.

Troubleshooting and measuring noise performance

Measuring noise in and around digital signals is difficult, if not impossible, without turning off adjacent signals. Digital signals are like noise, and occupy a significant portion of the spectrum so measuring noise ratios is difficult unless there are empty adjacent channels. This is discussed in "Intermodulation distortion" on page 212. The C/N in and around an analog television channel can be measured on the upper sideband side, or in the video sideband frequency range using non-interfering cable television analyzer techniques when blank interval test lines are available.

Reference

Analog

Digital

Figure 11.5. Picture quality degradation caused by noise.

Excessive noise causes picture to break out in blocks

The effects of noise on analog signals are well known. Snow and faded color of the image become more pronounced as the C/N is smaller. This is represented by the drawings in Figure 11.5. A digital video signal often responds to poor noise ratio by being dotted with small monochrome blocks. A change in the noise ratio beyond the threshold of a few tenths of a decibel causes the data stream to have such a high error rate that the digital demodulator loses synchronization, and the picture becomes unrecoverable, that is, over the cliff. Noise might not show up with this particular effect. A digital video picture's reaction to noise is dependent upon modulation format, error correction, adaptive equalization, and video compression.

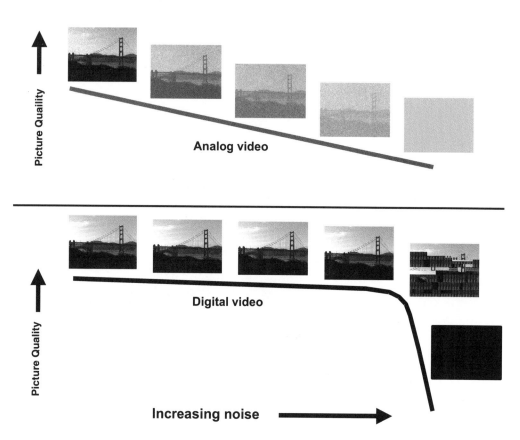

Figure 11.6. Picture quality response to system noise.

Figure 11.6 uses graphic illustrations to demonstrate how analog and digital video pictures respond to increasing noise in the system. The analog signal degrades according to C/N; the digital signal stays clear until near the cliff, then crashes in just 1 or 2 dB of increasing noise.

It is important for your system technical maintenance procedures to include guidelines on what to look for on which channels. Watching a picture, or listening to a subscriber complain about their picture is the last way you want to know your system is failing. Adding digital services which require quality data transfer in both directions must be maintained so that subscribers don't have an opportunity to complain. If they do you run the risk of losing their business. Your testing must include the measurement of system margin in some form.

Noise/distortion effects are different for different types of modulation

Use noise power ratio for digital signal-to-noise measurements

Different modulation formats are affected by noise and distortion differently. As the modulation gets more complex, that is, the number of symbols increases, the transmission efficiency goes up, but so does vulnerability to noise and distortion. For example, 16-QAM can carry a higher payload than QPSK, but it needs 8 to 12 dB better signal-to-noise ratio, depending upon error correction methods.

A traditional measurement in digital telecommunications is the comparison of the signal level with the level at a frequency purposely kept free of transmitted noise. A notch filter at the source provides the noise-free spectral space. Since the measurement is a comparison of noise levels, it is called **noise power ratio,** or **NPR**. It provides a graphical view of the signal-to-system-noise ratio. These are the conditions necessary to make accurate NPR measurements:

1) the adjacent signals are digital

2) the noise notch is free of composite or coherent distortion

3) the measurement instrumentation has sufficient dynamic range to see the full range of noise accurately

The idea is to use a notch that is either indigenous to the system, that is, a space between digital channels, or create a notch in a test channel, that is, one which is out of service and filled with wideband noise. The use of the space between channels is similar to using the lower channel boundary between analog television signals to look at system noise, as illustrated in Figure 11.7. If a notch is not available, one can be provided at the head end or source of the signals by inserting a narrowband notch filter whose rejection is 45 to 50 dB at the center frequency. Using a notch filter guarantees that the noise which fills in the notch is generated by system noise and is not channel sidebands.

The technique also assumes that the energy that fills the notch is random noise, and not CSO- or CTB-like semi-coherent distortion. The measurement rules for coherent distortion require the comparison of absolute powers. NPR is measured with the same detection receiver parameters. This condition leads to the third point: the measurement receiver, whether it is a spectrum analyzer, cable TV analyzer, or signal level meter, must measure the plateau signal as well as the valley with equal accuracy. For a spectrum analyzer to make this measurement, it is critical not to overload its input. Use a tunable channel filter centered at the notch. This prevents measurement error from analyzer mixer compression. The analyzer's input attenuator must be set to zero in order to maximize its sensitivity to the low notch noise.

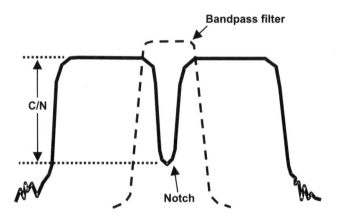

Figure 11.7. Using a noise notch to measure signal-to-noise.

The spectrum analyzer can also be used to quickly spot excessive noise by looking at the unfiltered notches between channels. In a quick look with an analyzer the notch should be more than 30 dB down. A notch of 40 dB is excellent because under these rough measurement conditions the spectrum analyzer tends to display signal-to-noise worse than it is. If you measure less than 30 dB without a bandpass filter, try it with the filter to get rid of any analyzer error.

Conditioning for cable channels

Telecommunications and telephony communication lines, such as twisted pairs, are rated according to their capabilities to carry signal reliably at specified rates. As cable television systems become cable telecommunications systems, they too have to have standards established and reinforced to compete with other common carriers of telephony and commercial data. A significant part of conditioning standards includes bit rates, attenuation and group delay, as well as signal-to-noise limits.

Careful system design and test procedures permits margin to be measured using signal-to-noise. See "MER diagnostics for troubleshooting and margin" on page 146. A system design that gives you the plot of Figure 11.8 for each major branch of your system, would make it possible for you to use NPR, or some other trustworthy technique for measuring signal-to-noise, to predict BER and margin.

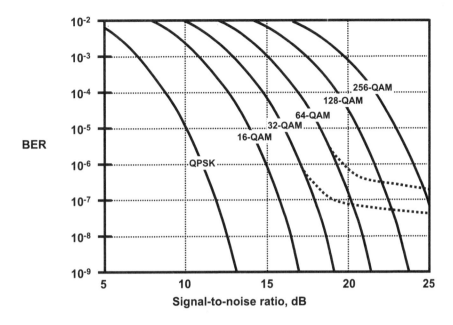

Figure 11.8. BER versus signal-to-noise ratio for various modulation formats.

Unfortunately, the parameters of this plot do not consider realistic systems, where composite intermodulation noise, or CIN, is the predominate noise base in the system. Figure 11.8 is drawn for noise signals in noise background environments. The theorists like it because it explains the way nature works. But CIN, is built into the nature of today's cable and HFC systems. The effects of CIN on these curves would be to flatten each modulation's performance curve to the right, limiting its improvements to some BER floor. This is illustrated in the figure with the dotted lines. (These dotted lines are conceptual. There is no experimental or formula-based data for their plot.) Curves of system performance must be based on actual performance, not theory. Practical measurements of modulation quality, EVM, MER, error rates of dedicated subsystems, and analog testing are the only sure ways to establish and maintain system quality.

Interference

There are no shortcuts to minimizing interference. An ingress-free and corrosion-proof signal path from the head end through the customer home is necessary to maintain reliable digital transmission.

The case for quality connections cannot be stated too strongly for systems relying upon the return path for data transmission.

Interference, whether it comes from within the system, or from ingress from outside the system, must be identified before it can be fixed. In the RF domain, the spectrum analyzer is a powerful tool to observe ingress and identify its modulation profile, transmission frequency, and/or magnitude. In the baseband domain, the use of a modulation analyzer to observe the symbol constellations can also help identify the type of disturbance. CW and modulated signals are easy to identify by observing the constellation or demodulating the interfering signal, as shown in "Troubleshooting with constellation diagrams" on page 135.

Test equipment for noise and distortion measurements

A host of test equipment make measurements related to distortion, noise, and interference. Figure 11.9 shows the influences of these parameters and the measurement equipment that makes the measurements.

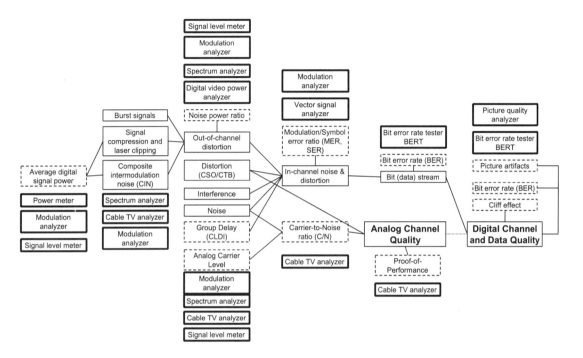

Figure 11.9. The measurement equipment used to make distortion, noise, and interference measurements.

Summary

Distortion, noise, and interference affect digital signals by eroding the digital information encoded inside them until the data corrections can no longer sustain their high quality, and it goes over the cliff. Without analog signal quality, it is nearly impossible to troubleshoot and diagnose digital signal quality problems.

The system performance that causes distortion, noise and interference are generally the same as those that cause analog signal problems: intermodulation, signal compression, system and in-channel noise, transmission group delay, and ingress.

Intermodulation distortion creates unwanted signals from the mixing of desired and undesired signals, which is noise-like. This composite intermodulation noise, or CIN, appears at the frequencies throughout the channels in a random fashion, rather than in the more statistically predictable places CSO and CTB fall because of the frequency placement differences between analog carriers and wide digital channels.

Digital signals are susceptible to distortion no matter where the distortion products lie within their channel. PAL or NTSC signals are less affected to distortion products which fall in the channel band, but outside the specified video and audio ranges. Signal-to-noise ratio measurements require adjacent empty channels, and in-channel effects cannot be measured with non-interfering techniques similar to those used in C/N for analog signals today.

Coherent distortion within the channel bandwidth cause BER and SER increases; distortion close to the channel may or may not affect the margin depending upon the amount of signal data compression, its symbol rate, and its modulation complexity. There is practical evidence that suggest a -6 to -10 dBc ratio between the analog video carrier peak and the digital video average power minimizes the effects of disruptive interaction between them.

Carrier-to-noise ratio, traditionally held as the single most important measure of system performance, is going to yield to MER as HFC and cable systems are filled with digital signals.

The measurement of noise level becomes more important as your system begins to fill up with digital signals because they add CIN. Measuring noise in and around digital signals is difficult, if not impossible, without turning off adjacent signals. A notch filter at the source provides the noise-free spectral space by which to measure the noise generated by CIN and system noise. This technique is called noise power ratio, or NPR. A notch in the noise is created at the head end. How much noise fills the notch in the forward path indicates the system noise conditions. The spectrum analyzer can be used to quickly spot excessive noise by looking at the unfiltered notches between channels. In a quick look with an analyzer the notch should be more than 30 dB down. Careful system design and test procedures permit margin to be measured using signal-to-noise.

But the heart of monitoring your system performance are the non-interfering measurements you can make conveniently in the head end, nodes, and in the field. These include EVM, MER, error rates of dedicated subsystems, and analog testing. And always start with the best analog performance system you can afford.

There are no shortcuts to minimizing interference. An ingress-free and corrosion-proof signal path from the head end through the customer home is necessary to maintain reliable digital transmission. Interference, whether it comes from within the system, or from ingress from outside the system, must be identified before it can be fixed.

Questions for review

1. What is a minimum performance requirement for protecting digital signals from distortion, noise, and interference? Choose all that apply.
   a) Conformance to analog signal proof-of-performance standards
   b) Minimized ingress and signal path corrosion
   c) High quality of the signals at the head end

2. The generally accepted power ratio between adjacent digital signals and an analog PAL or NTSC carriers is
   a) 0 to 4 dB, where the analog signal is stronger
   b) 4 to 8 dB, where the analog signal is weaker
   c) 8 to 12 dB, where the analog signal is stronger
   d) ingress-free and corrosion-proof signal path from head end to subscriber.
   e) none of these

3. The distortion produced by digital signals is noise-like. True or False?

4. Composite intermodulation noise, CIN, is
   a) the distortion caused by mixing of digital signals in the system's amplifiers
   b) just like the CW characteristics of the CSO and CTB from analog channels
   c) likely to appear anywhere in the system's frequency range
   d) often indistinguishable from system noise
   e) all of the above

5. Signal-to-noise ratio for digital signals (select all that are true)
   a) can always be used to predict a digital signal's margin
   b) can be used to predict a digital signal's margin if the system and measurement equipment has been designed for that purpose
   c) can be measured using the noise power ratio, NPR, method
   d) can be estimated using a spectrum analyzer
   e) all of the above

6. Digital signals generally are more tolerant of system noise and CIN than analog signals. True or False?

7. The best guard against interference is corrosion-proof connections and ingress-free transmission path.

Selected bibliography

1. *Interval*, the SCTE Newsletter, Society of Cable Telecommunications Engineers, Inc. *DigiPoints*, June 1997.
2. Edwin Cooper and Neil Abramson, "A Brave New Digital World," *Communications Technology* magazine, Communications Technology Publications, Inc., October 1997.
3. John Ernandez, "Understanding the Effects of Noise on Digital Signals," *Communications Engineering & Design*, Chilton Publications, New York, December 1997.
4. Joe Gorin, *Spectrum Analyzer Measurements and Noise*, Application Note 1303, Hewlett-Packard Company, publication number 5966-4008E, May 1998.
5. Ron Hranac, "Learn How to Make Two-Way Work," *Communications Technology* magazine, Communications Technology Publications, Inc., October 1997.
6. M. Stephen McConnell, "Effects of Analog and Digital Signals," *Communications Engineering & Design*, Chilton Publications, New York, December 1996.
7. Kenneth H. Metz, "Going Digital? Think Bit Error Rate", *Communications Technology* magazine, Communications Technology Publications, Inc., June 1997.
8. Donald Raskin and Dean Stoneback, *Broadband Return System for Hybrid Fiber/Coax Cable TV Networks*, ISBN 0-13-636515-9, Prentice Hall PTR, Upper Saddle River NJ, 1998.
9. Matt Trezise, "Understanding the Measures of Signal Quality in DVB Systems," paper associated with DVB-C 8594Q application training, Hewlett-Packard Company, 1995.
10. Jeffrey L. Thomas, *Cable Television Proof-of-Performance; A Practical Guide to Cable TV Compliance Measurements Using a Spectrum Analyzer*, ISBN 3-13-306382-8, Hewlett-Packard Press, Prentice Hall PTR, Upper Saddle River NJ, 1995.
11. Andrew S. Tanenbaum, *Computer Networks*, 3rd Edition, ISBN 0-13-349945-6, Prentice-Hall, Inc., 1996.
12. Ernest Tsui and Michael Meschke, "Digital testing: QAM fundamentals," *Communications Technology* magazine, Phillips Business Information Inc., Rockville MD, May 1996.

Appendix A - Glossary

Terms and multiple-letter acronyms used in this book:

- A -

ACP. *See* adjacent channel power

adaptive equalization. A signal processing function in a receiver that removes linear distortion due to the transportation media.

adjacent channel. The channel (frequency band) immediately above or below the channel of interest.

adjacent channel power (ACP). A comparison between the channel powers to either side of the channel under test, usually given as a power ratio in dB.

AE. *See* adaptive equalization

aliasing. When converting a signal from analog to digital, aliasing misinterprets the sampled analog signal such that the original signal cannot be restored from the digital data.

AM. *See* amplitude modulation

amplifier. Device used to increase the operating level of an input signal. Used in a cable system's distribution plant to compensate for the effects of attenuation caused by coaxial cable and passive device losses.

amplitude. The size or magnitude of a voltage or current waveform; the strength of a signal.

amplitude modulation. The form of modulation in which the amplitude of the signal is varied in accordance with the instantaneous value of the modulating signal.

analog. For purposes of communication, analog is the non-quantized storage and transmission of data.

attenuator. A device for reducing the amplitude of a signal.

analog to digital conversion (ADC). Pertaining to signals in the form of continuously variable physical quantities.

attenuation. The difference between transmitted and received power due to loss through equipment, lines, or other transmission devices; usually expressed in decibels. In spectrum analyzer operation, a general term used to denote a decrease of signal magnitude in transmission from one point to another. Attenuation may be expressed as a scalar ratio of the input to the output magnitude in decibels.

artifact. In the subjective observation of video, unwanted picture effects due to the discarding of visual detail by processes usually related to signal compression.

ADSL. *See* digital subscriber line

asynchronous transmission. A method of sending data over a communications line by placing a block of transmitted bits in an envelope. The envelope begins with a start bit that tells a computer a character is beginning. The stop bit sends a message that a character has ended. Asynchronous transmission also has the advantage of not needing precise clocking mechanisms that maintain a time relationship between transmitter and receiver.

asynchronous transfer mode (ATM). An international packet switching standard using a cell-switched approach in which each packet of information features a uniform size of 53 bytes. Of the total cell, 48 bytes is the payload or information to be transferred. Five bytes are used as a header, providing all the addressing information for that particular packet. ATM can switch and route information of all types; video, voice, and data simultaneously.

ATM. *See* asynchronous transfer mode

ATV. advanced television, sometimes called DTV for digital television

aural carrier. The carrier that has the audio portion of a television channel. A television channel usually contains both a visual and an aural carrier. An aural carrier is sometimes referred to as a sound carrier.

average power. *See* power level

- B -

bandpass filter. A device that allows signal passage to frequencies within its design range and that effectively bars passage to all signals outside that frequency range.

bandwidth. (1) A measure of the information carrying capacity of a communication channel. The bandwidth corresponds to the difference between the lowest and highest frequency signal that can be carried by the channel. (2) The range of usable frequencies that can be carried by a cable television system.

baseband. The band of frequencies occupied by the signal in a carrier wire or radio transmission system before it modulates the carrier frequency to form the transmitted line or radio signal.

baseband channel. Connotes that modulation is used in the structure of the channel, as in a carrier system. The usual consequence is phase or frequency offset. The simplest example is a pair of wires that transmits direct current and has no impairments such as phase offset or frequency offset that would destroy waveform.

baseband signal. A signal transmitted at the frequencies in which it was first created.

baud. The number of times a state change occurs on a communications channel per second. Baud is not bit rate, or bits per second. The baud rate equals the bit rate only when the state change represents a single bit of data, which depends upon the symbol definition of the transmission.

BER. *See* bit error rate

BERT. *See* bit error rate tester

BFSK. binary frequency shift keying

bit error rate (BER). A measure of transmission accuracy. It is the ratio of bits received in error to the bits sent. Also called bit error ratio.

bit error rate tester (BERT). A measurement instrument for monitoring the bit error rate of a data stream either by comparing known bit streams to the actual bit streams or by deducing the error rate from the rate at which the receiver is applying error correction.

bit (b). A one or a zero representing a single value in a binary counting system. Bit is a contraction from the words binary digit.

bit rate . The speed at which digital signals are transmitted, expressed in bits per second, bps.

block. A specific amount of digital information used to define part of an MPEG2 picture frame.

broadband (BB in RF cable networks). Any communications system able to deliver multiple channels or services of video, voice, or data to its users or subscribers using RF spectrum.

broadband (data transport networks). Broadband is any transmission facility that has a bandwidth (capacity) greater than a voice-grade line of 3 kHz.

broadcast. Information distribution in which the data are distributed to all participants of the system, but only used by the intended receivers.

burst. In communications, a burst signal is one with non-continuous power changes that dip close to zero power.

BW. *See* bandwidth

byte. A digital word of 8 bits.

- C -

C/N margin. The difference between the current carrier-to-noise ratio of a system from the ratio that cause signal quality problems.

C/N. *See* carrier-to-noise ratio

cable television. A broadband communications technology in which multiple television channels as well as audio and data signals are transmitted either one way or bidirectionally through a distribution system to single or multiple specified locations. The term also encompasses the associated and evolving programming and information resources that have been and are being developed at the local, regional, and national levels.

cable television system. A broadband communications system capable of delivering multiple channels of entertainment programming and non-entertainment information from a set of centralized antennas, generally by coaxial cable, to a community. Many cable television designs integrate microwave and satellite links into their overall design, and some now include optical fibers as well. Often referred to as cable television, which usually stands for community antenna television system.

carrier. An electromagnetic wave, of which some characteristic is varied to convey information.

carrier-to-noise ratio (C/N). The ratio of amplitude of the carrier to the noise power relative to a 4-MHz bandwidth in the portion of the spectrum occupied by the carrier. Also referred to as the C/N ratio, or C/N.

carrier-to-noise ratio (C/N). For analog cable television signals, C/N is the ratio of the amplitude of the carrier to the noise power relative to a 4-MHz bandwidth in the portion of the spectrum occupied by the carrier.

CATV. Abbreviation for community antenna television or cable television system. A cable television system is a broadband communications system that provides multiple channels from centralized antennas.

CCITT. Comitre Consultatif Internationale de Telegraphique et honique. An international group operating under the auspices of the International Telecommunications Union and charged with establishing telecommunications standards. The name was changed to ITU-TSS (International Telecommunications Union-Telecommunications Standards Sector).

CDMA. code division multiple access

cell relay. A packet switching technique that uses cells of uniform length. Cell relay is well suited to video transmissions where the predictable arrival of packets in proper order is important.

center frequency. (1) The average frequency of the emitted wave when modulated by a sinusoidal wave. (2) The frequency of the emitted wave without modulation.channel A signal path of specified bandwidth for conveying information.

channel. A signal path of specified bandwidth for conveying information.

channel spacing. The difference frequency, in Hz, between two adjacent channel carriers, provided the carriers are located in the same relative places within the respective channels. *See* channel bandwidth.

channel bandwidth. The transmission bandwidth, in Hz, required to transport information in a channel. Channel bandwidth may or may not equal the channel spacing.

CIN. *See* composite intermodulation noise

cliff effect. Also called the waterfall effect. The cliff effect describes the way a digital signal usually fails when overwhelmed by transport impairments. It happens when the receiver can no longer error correct or equalize out the impairments. Unlike analog signals which degrade with a whimper, digital signals die with a bang.

cliff effect curve. The plot of symbol error rate against carrier-to-noise ratio which predicts the tolerance of a signal to additional system noise. Also *see* cliff effect.

clipping (of return laser). A distortion caused by overloading a laser amplifier. As the input signal grows in amplitude, the lower side is clipped off by the non-linear portion of the power transfer curve, giving rise to large amount of distortion.

coaxial cable. A type of cable used for broadband data and cable systems. Composed of a center conductor, insulating dielectric, conductive shield, and optional protective covering, this type of cable has excellent broadband frequency characteristics, noise immunity, and physical durability. Synonymous with coax.

code division multiplex (CDMA). A transmission scheme used by some cellular carriers derived from spread spectrum techniques used by the military.

COFDM. coded orthogonal frequency division multiplex

composite. The effect of several distortion signals present within a very narrow bandwidth. *See* discrete.

composite intermodulation noise (CIN). The distortion energy created by intermodulation between two or more digital signals, which appears as noise.

composite second-order beat (CSO). In NTSC video channels, (1) A clustering of second order beats 1.25 MHz above the visual carriers in cable systems. (2) A ratio, expressed in decibels, of the peak level of the visual carrier to the peak of the average level of the cluster of second-order distortion products located 1.25 MHz above the visual carrier.

complex. The treatment of signals, voltages, currents, and impedances as composed of interconnected parts. A vector signal is one which is specified fully only if its magnitude and angle are give.

compression, power or gain. (1) A less-than-proportional change in output for a change in input. (2) The reduction in amplitude of one portion of a waveform relative to another portion.

compression, data. In the transmission of video signals, data compression is a digital process that reduces the volume of data by removing redundant adjacent pixel information in a single TV frame and redundant information from frame to frame.

computer network. The transmission and encoding system that facilitate computers to communicate with each other.

constellation diagram. A constellation diagram is a from of vector diagram that plots the symbol points on an I/Q field.

converter. Also known as processor. Device for changing the frequency of a television signal. A cable head end converter changes signals from frequencies at which they are broadcast to clear channels that are available on the cable distribution system. A set-top converter is added in front of a subscriber's television receiver to change the frequency of the midband, superband, or hyperband signals to a suitable channel or channels (typically a low- VHF channel) which the television receiver is able to tune.

cross compression. Effect of ingress on return path active device; one impulse at one frequency compresses all the signals in the device's bandwidth.

cross modulation. A form of television signal distortion where modulation from one or more television channels is imposed on another channel or channels.

CW. continuous wave

- D -

DAB. digital audio broadcast

data error. An error in a digital signal transmission system that causes information to be changed or lost in transit.

data stream. A collection of characters and data bits transmitted through a channel.

dB. *See* decibel

dBm. *See* decibel milliwatt

dBmV. *See* decibel milliwatt

decibel (dB). A unit that expresses the ratio of two power levels on a logarithmic scale.

decibel millivolt (dBmV). A unit of measurement referenced to one millivolt across a specified impedance.

decibel milliwatt (dBm). A unit of measurement referenced to one milliwatt across a specified impedance.

demodulate. To retrieve an information-carrying signal from a modulated carrier.

DBS. direct broadcast satellite

diagram. *See* I/Q diagram, polar diagram, vector diagram

digital modulation. *See* modulation

digital. For the purposes of communications, digital is the quantized storage and transmission of data, that is the use of a binary code to represent information.

digital converter. An electronic device that converters a baseband or digitally modulated RF signal to a useful form. A set-top box is a digital converter for cable systems with digital subscriber services.

digital modulator. *See* digital modulator

digital signal. A discontinuous signal made up of discrete values of amplitude, frequency or phase that represent information.

digital signal processing (DSP). The conversion of an electronic signal from one format or style to another for more effective use of its information.

digital subscriber line (DSL). High data rate communications line to home, usually provided by the telephone companies. Asymmetric digital subscriber line (ADSL) converts existing twisted-pair telephone lines into access paths for high-speed communications.

digital video broadcast (DVB).

digitize or digitalization. The process of converting information into digital data.

digitizing center. A system that digitizes signals at a central location and distributes them to remote head ends. Also called a digital turnaround system.

discrete. Distortion and interference signals generated from a combination of known tones. Also *see* composite.

distortion. An undesired change in waveform of a signal in the course of its passage through a transmission system.

distribution system. The part of a cable television system consisting of trunk and feeder cables that are used to carry signals from the system head end to subscriber terminals. Often applied, more narrowly, to the part of a cable television system starting at the bridger, amplifiers. Synonymous with trunk and feeder system.

DOCSIS. Data-Over-Cable Service Interface Specification. In North America, a group of cable system operators, in a consortium called the MCNS, are creating a standard for a data over cable system which adds to a cable television system a high- speed data communications path that is transparent to the Internet Protocol (IP), between subscriber locations and the cable operator's headend. CableLabs (tm) is tasked with administering the specification process. With the assistance of the Society of Cable Telecommunication Engineers (SCTE), intend to submit the standard to the American National Standards Institute (ANSI) and the International Telecommunications Union (ITU).

DQPSK. differential quadrature phase shift keying

DSL. *See* digital subscriber line

DSP. *See* digital signal processing

DVB-C. digital video broadcast for cable

DVB-S. digital video broadcast for satellite

DVB-T. digital video broadcast for terrestrial

dynamic range. In general, the ratio (in decibels) of the weakest or faintest signals to the strongest or loudest signals reproduced without significant noise or distortion. In a spectrum analyzer, the power ratio (dB) between the smallest and largest signals simultaneously present at the input that can be measured with some degree of accuracy. Dynamic range generally refers to measurement of distortion or intermodulation products. Also, the operating amplitude power range for an RF or laser amplifier. Dynamic range is also an important parameter in measurement instruments, such as, spectrum analyzers, and TDRs.

- E -

echo. An echo is the reflection of electromagnetic energy in a transport medium such as cable caused by a discontinuity in the medium's impedance.

encoding. The process of converting data into digital code.

Ethernet. A data communications protocol operating at up to 10 Mbps.

ETSI. European Telecommunications Standardization Institute

EVM. *See* error vector magnitude

eye diagram. An eye diagram is a side view of the constellation vector trail diagram, either in the I-plain or the Q-plain. Eye diagrams show how the I and Q vector trails behave over time.

- F -

fast Fourier transform (FFT). A mathematical operation performed on a time-domain signal to yield the individual spectral components that constitute the signal.

FDM. *See* frequency division multiplexing

FDMA. *See* frequency division multiple access

FEC. *See* forward error correction

FFT. *See* fast Fourier transform

fiber-optics. The rapid transmission of light pulses in a coded digital format through the fiber cable. In a fiber-optic transmitter, a light source such as a laser or light-emitting diode (LED) is connected to the fiber cable. This light source converts an electronic input signal into a series of light pulses by blinking on and off millions of times per second. This stream of light pulses is the combination of many lower rate bit streams formed using digital multiplexing techniques. At the other end of the fiber, receivers capture the light pulses for conversion to electrical signals.

field. One-half of a complete picture (or frame) interval, containing all of the odd or even scanning lines of the picture.

field strength. The intensity of an electromagnetic field at a given point, usually referred to in microvolts per meter.

filter. A filter is a passive device that changes the bandwidth of a signal by limiting its range and/or changing its shape.

flatness. *See* frequency response

FM. *See* frequency modulation

format. *See* modulation format

forward error correction (FEC). Forward error correction is an encoding and decoding process where redundant data are sent along with the signal to allow their repair at the receiver if the receiver detects that the original information has been corrupted.

forward path. The transmission from the head end to the subscriber. Also called downstream.

Fourier transform. *See* fast Fourier transform

frame. In digital signal transport, a frame is a subset of the data stream defined by the signal's format which contains the information necessary for its control, identification, and addressing of the signal data, or payload, within the frame.

frame relay. A data transmission technique that features the use of a flag as a start bit, an address to indicate intended destination of the message, the actual payload to be delivered, an error detection sequence and a flag indicating end of the packet.

freeze frame. The stopping of a video picture long enough for the viewer to notice.

frequency. The rate at which electromagnetic oscillates in Hertz, or Hz, in units of cycles per second.

frequency domain. A view of a signal's behavior as a function of frequency.

frequency modulation (FM). A form of modulation in which the frequency of the carrier is varied in accordance with the instantaneous value of the modulating signal.

frequency response. (1) The relationship within a cable television channel between amplitude and frequency of a constant amplitude input signal as measured at a subscriber terminal. (2) The measure of amplitude frequency distortion within a specified bandwidth.

frequency stability. Stability is the ability of a frequency component to remain unchanged in frequency or amplitude over short- and long-term periods of time. In spectrum analyzers, stability refers to the local oscillator's ability to remain fixed at a particular frequency over time. The sweep ramp that tunes the local oscillator influences where a signal appears on the display. Any long-term variation in local oscillator frequency (drift) with respect to the sweep ramp causes a signal to shift its horizontal position on the display slowly. Shorter-term local oscillator instability can appear as random FM or phase noise on an otherwise stable signal.

FSK. frequency shift keying

- G -

Gbps. giga bits per second. A data rate of 1 billion (1,000,000,000) bits transmitted every second.

giga. Multiply by 1,000,000,000.

group delay. The amount of time it takes a signal to get through a system. Linear group delay means that each signal within the system's frequency range has the same group delay. Non-linear group delay means that a signal's time through the system varies with the signal's frequency.

- H -

harmonic distortion. (1) The generation of harmonics by the circuit or device with which the signal is processed. (2) Unwanted harmonic components of a signal.

HDTV. high definition television

head end. The control center of a cable television system, where incoming signals are amplified, converted, processed, and combined into a common cable, along with any origination cablecasting, for transmission to subscribers. System usually includes antennas, preamplifiers, frequency converters, demodulators, modulators, processors, and other related equipment.

Hertz (Hz). A unit of frequency equivalent to one cycle per second.

heterodyne. To mix two frequencies together in a non-linear component to produce two other frequencies equal to the sum and difference of the first two. Synonymous with beat.

heterodyne processor. An electronic device used in cable head ends that down-converts an incoming signal to an intermediate frequency for filtering, signal level control, and other processing, and then reconverts that signal to a desired output frequency.

HFC. hybrid fiber coax

hum distortion. Undesired modulation of the television visual carrier by power-line frequencies or their harmonics (e.g.,50-60 or 100-120 Hz), or other low-frequency disturbances.

Hz. *See* Hertz

- I -

I. incidental or in-phase

I/Q. incidental or in-phase/quadrature

ISI. *See* intersymbol interference

IEEE. Institute of Electrical and Electronic Engineers

IF. intermediate frequency

impairment. In an RF transport medium, such as cable, an impairment is any thing that causes degradation to a transported signal's quality. Impairments can be physical, such as a broken connector, or electronic, such as distortion or ingress.

incidental (I). In definition of a vector signal, it is incidental if it is aligned with the carrier's angle. Also called in-phase. Also *see* quadrature.

impedance. The combined effect of resistance, inductive reactance, and capacitive reactance on a signal at a particular frequency. In cable television, the nominal impedance of the cable and components is 50 or 75-Ohms.

in-channel. A characteristic or event that occurs within the transmission channel bandwidth. For example, intermodulation distortion falling within its own transmission bandwidth is an in-channel distortion.

interlacing. In video formats, interlacing is a technique for using every other horizontal line for updating the frame. Alternate lines are updated each frame, saving transmission bandwidth for the number of horizontal lines shown.

intersymbol interference. symbols misread by the demodulator

in-service. In-service usually refers to the ability to do maintenance and repair on system facilities without disrupting service to the subscribers.

interoperability. The level at which system components from different manufacturers are able to work together.

information. Message content that cannot be anticipated by the recipient. (such as your next power bill) In information theory, information content is zero if the whole message is known in advance, to 100 percent if the message is completely unknown in advance of its reception.

ingress. The unwanted leakage of interfering signals into a cable television system.

input attenuator. An attenuator (also called an RF attenuator) between the input connector and the first mixer of a spectrum analyzer. The input attenuator is used to adjust the signal level incident to the first mixer, and to prevent gain compression due to high-level or broadband signals. It is also used to set the dynamic range by controlling the degree of internally generated distortion. For some spectrum analyzers, changing the input attenuator settings changes the vertical position of the signal on the display, which then changes the reference level accordingly.

input impedance. The terminating impedance that the spectrum analyzer presents to the signal source. The nominal impedance for RF and microwave spectrum analyzers is usually 50-Ohms. For some systems, such as cable TV, 75-Ohms is standard. The degree of mismatch between the nominal and actual input impedance is called the VSWR (voltage standing wave ratio).

insertion loss. Additional loss in a system when a device such as a directional coupler is inserted, equal to the difference in signal level between input and output of such a device.

intersymbol interference (ISI). In data transmission, intersymbol interference is the distortion of signals due to preceding or following pulses affecting the desired pulse amplitude at the time of sampling.

IP. internet protocol

interference, electromagnetic. Any electromagnetic energy, natural or man-made, which may adversely affect performance of the system.

ISDN. integrated services digital network

- J -

jitter. short-term variations of the significant instance of a digital signal from their ideal position in time

JPEG. A data compression format used primarily for still pictures.

- K -

kbps. kilo bits per second, or 1000 bits transmitted per second

kilo. Multiply by 1000

- L -

LAN. *See* local area network

layer, signal. A signal layer is one of several encoding processes designed to keep its information compact and protected from transmission problems. When a digital signal is sent over a computer network, the protected digital signal, already layered within itself, is processed through the computer network's layers.

layer, computer network. Layer in a computer network are encoding processes to prepare a signal for multiplexing with others and for reliable and measurable data flow transport on the physical layer, the layer that actually transmits the data.

leakage. Undesired emission of signals out of a cable television system, generally through cracks in the cable, corroded or loose connections, or loose device closures. Synonymous with signal leakage.

linear. The power into a device or circuit is proportional to the power out of the device or circuit. Also *see* non-linear.

line error. Conditions in the transmission medium that cause errors in the digital signals being sent. Line errors defined for twisted pair for telephony include attenuation distortion, envelope delay distortion, signal-to-noise ratio, harmonic distortion, impulse noise, frequency shift, echo, and jitter.

local area network (LAN). A limited distance network connecting a defined set of terminals. For example a LAN can be used to connect workstations within an office, offices in a building, or buildings within a campus.

local oscillator (LO). An oscillator, built into the design of the equipment, that generates a signal used in the heterodyne process to mix with incoming signals and produce an intermediate frequency.

low pass filter (LPF). A filter which passes all frequencies below a specified frequency, and blocks those frequencies above the specified frequency.

- M -

magnitude. The amplitude of a signal in terms of its absolute value, stripped of any plus or minus sign.

mapping. The definition of bit combinations in a constellation diagram which define symbols.

match, impedance. A method used to match two or more components into a single characteristic impedance of one of the components, to minimize attenuation and anomalies.

margin. An informal term to describe how close to the cliff edge a signal or system is operating. There are many well-defined parameters for this measurement, including, margin-to-critical BER, C/N margin and modulation error ratio (MER).

mega. multiply by 1,000,000

MER. *See* modulation error ratio

microreflections. *See* structural return loss

microsecond (μs or μsecond). One millionth of a second

microwave (mw). A very short wavelength electromagnetic wave, generally above 1000 MHz.

mismatch. (1)The condition resulting from connecting two circuits or connecting a line to a circuit in which the two impedances are different. (2) Impedance discontinuity.

MMDS. multichannel multipoint distribution service

modulation. The process of adding information to a signal. If the signal is electrical, the information is added to the signal by varying one or more of the signal's characteristics, such as its amplitude, frequency or phase. Digital modulation means that the information added to the carrier is digital in nature, that is, it is comprised of a bit stream without regard to the information carried.

modulation error ratio (MER). Modulation error ratio compares the modulation error power to the average transmission power of the channel.

modulation format. A name given to a specific modulation type, such as, amplitude modulation (AM), quadrature amplitude modulation (QAM), and coded orthogonal frequency division multiplex (COFDM).

modulation quality. Modulation quality is a measure of how well a signal survives the modulation/demodulation processes through a transport system. The measure is usually a number derived from the comparison of a transmitted signal compared to an ideal signal.

mosaic. video effect

MPEG. Moving Picture Experts Group, an international standards-setting group working under the supervision of the International Standards Organization and the International Electrotechnical Commission to develop standards for compressed full-motion video, still image, audio and other associated information. Coding and decoding audio, and video, and the transport and storage of video are the primary concerns of the MPEG.

multimedia. The integration of at least two of five information type for presentation on a TV set, desktop computer screen, personal information manager or other computer-driven device with a screen interface. Multimedia information can include text still graphics, animation, audio, full-motion video, or still photos.

multiplexing. Any technique that permits simultaneous transmission of multiple signals over one circuit.

- N -

narrow band. A relative term referring to a system that carries a narrow-frequency range (sometimes used to refer to frequency bandwidths below 1 MHz). In a telephone/television context, telephone would be considered narrow band (3 kHz) and television would be considered broadband (6 to 8 MHz).

network. *See* computer network

noise. That combination of undesired and fluctuating disturbances within a cable television channel that degrades the transmission of the desired signal. Also, random burst of electrical energy or interference which may produce a "salt-and-pepper" pattern over a television picture. Heavy noise is sometimes called "snow."

noise figure. The amount of noise added by signal-handling equipment (e.g., an amplifier) to the noise existing at its input, usually expressed in decibels.

noise power ratio (NPR). A technique for measuring the amount of composite intermodulation noise in a system, where the intermodulation is primarily from the mixing of digital signal. A noise notch at the system head end is monitored along the forward path. The more the notch fills in, the worse the composite intermodulation noise.

non-intrusive. A measurement is non-intrusive if it does not disrupt service.

non-linear. Power out from a system is not proportional to the power input. Also *see* linear.

NTSC. National Television System Committee

Nyquist criteria. The Nyquist criteria state that a sampling rate of two times or greater than the highest frequency of the signal is required to preserve the waveform frequency in the quantized data.

- O -

OQPSK. offset quadrature phase shift keying

oscilloscope. An oscillograph test apparatus primarily intended to visually represent test or troubleshooting voltages with respect to time. Synonymous with scope.

out-of-channel. A characteristic or event that occurs outside the transmission channel bandwidth. For example, intermodulation distortion of a digital signal falling outside its own transmission bandwidth and interfering with an adjacent channel is out-of-channel distortion.

- P -

packet. A bundle of data packaged for transmission over a network. Packets can be various lengths, ranging from about 40 bytes up to 32,000 bytes on the Internet, but typically about 1500 bytes in length.

packet error. A data group in a compression data stream that is faulty, causing a glitch or block error in the picture.

PAL. phase alternate line (television)

PCM. pulse coded modulation

parameter. A measurable aspect of a device, system, or signal which characterize the device, system, or signal for the purposes of comparison to others, or to a specification. Sometimes called a metric.

PCS. personal communication services

passband. The range of frequencies passed by a filter, amplifier, or electrical circuit.

passive, passive device. A device basically static in operation; that is, it is not capable of amplification or oscillation, and requires no power for its intended function. Examples include splitters, directional couplers, taps, and attenuators.

pass-through. Taking the incoming transport stream at a head end and remodulating it for use in the cable system.

payload. The information to be sent from the transmitter to the receiver, exclusive of error correction, protocol, training, or addressing.

peak power. *See* power level

peak-to-average ratio. The ratio of a signal's peak power excursions to the its average power. A CW signal has a peak-to-average ratio of 3 dB.

period. The period, in seconds, of a signal is the reciprocal of its frequency, in Hz.

periodic. A signal which repeats its waveform in time.

phase. The angle of a signal's vector compared to the carrier.

phase modulation. Information transmitted on a carrier by changing its phase according to the information's waveform.

pi/4QPSK. pi over 4 quadrature phase shift keying

picture artifacts. In digital video, artifacts are the unintended, and unwanted visual aberrations in the picture image.

PID. *See* program identifier

PM. *See* phase modulation

polar diagram. A polar diagram is the display of a signal's vector magnitude and phase, relative to the carrier vector.

power level. Peak: the power over a radio frequency cycle corresponding in amplitude to synchronizing peaks. Refers to television broadcast transmitters.

power density. *See* power level

PRBS. pseudo random binary sequence

preamplifier. A low-noise electronic device (usually installed near an antenna) designed to strengthen or boost a weak off-air signal to a level where it overcomes the antenna downlead loss and be sufficient to drive succeeding processors or amplifiers. In spectrum analyzers, an external, low-noise-figure amplifier that improves system spectrum analyzer (preamplifier/spectrum analyzer) sensitivity over that of the spectrum analyzer itself.

protocol. A set of rules about how computers are to act when talking to each other. Examples include Ethernet, IEEE 802.5 token ring, X Modem, and Kerwin.

progressive. In video formats, a progressive format forms the picture as a complete frame using all the pixels displayed, without interlacing.

program identifier. In a data stream which contains several programs multiplexed together, the packets of data are tagged with a code recognized by the receiver's decoder that assigns packets to the correct program.

PSD. power spectral density

pulse. A variation in the value of a quantity, short in relation to the time schedule of interest, with the final value being the same as the initial value.

- Q -

QAM. quadrature amplitude modulation

QPSK. quadrature phase shift keying

quadrature (Q). In definition of a vector signal, the vector is quadrature if it is 90 degrees from the carrier's angle. *See* incidental.

qualitative measurement. In measuring signal quality, qualitative testing lets you observe the digital signal in progress, watching for unusual patterns that may be indications of specific impairments.

quantitative measurement. In measuring signal quality, quantitative measurements provide a number that can be compared to a specification.

- R -

radio frequency (RF). An electromagnetic signal above the audio and below the infrared frequencies.

random noise. Thermal noise generated from electron motion within resistive elements of electronic equipment.

RBW. *See* resolution bandwidth

real-time. In the transportation of television signals, real-time means that the receiver is able to present the signal at the exact same rate as it is transmitted. Delay through the system, due to transmission propagation speeds and coding/decoding times, do not affect the real-time nature of the signal, only the rate of its delivery.

receiver. The recipient of a signal for the purpose of decoding the signal's information for immediate usage, relay, or storage. Sometimes called a sink.

repeater. A device that restores a signal in a transmission media such as cable to a desired amplitude level.

residual measurement. A measurement technique that uses the difference between two data.

resolution bandwidth (RBW). In a spectrum analyzer, the resolution bandwidth is the IF filter that determines the analyzer's ability to separate two equal-level signals, space by a frequency equal to, or greater than the filter's 3-dB bandwidth. For example, if the RBW is 10 kHz, the analyzer can select between equal-level signals spaced 10 kHz or wider apart.

return path. The transmission from the subscriber to the head end. Also called upstream, reverse, and reverse path.

return loss. Return loss is a measure of how much of a signal's power is reflected back from a load or transmission impairment. A good match between a source and a load has a very high return loss, meaning that the energy reflected back to the source is very small.

RF. *See* radio frequency

RMS. root mean square

- S -

sample rate. In the analog to digital conversion processing, the sample rate is the frequency with which the analog signal's waveform values are stored for processing into digital data.

sample, sampling. In analog-to-digital conversion processing, a sample is a measure of the analog signal's waveform amplitude at an instant in time. *See* sample rate.

scalar. The measure of an electronic signal without regard to phase or angle.

SECAM. sequential width memory (television)

second harmonic. In a complex wave, a signal component whose frequency is twice the fundamental, or original, frequency.

second-order beat . Even-order distortion product created by two signals mixing or beating against each other.

sidebands. Additional frequencies generated by the modulation process, which are related to the modulating signal and contain the modulating intelligence.

SDTV. *See* standard definition television

SER. *See* symbol error ratio

server. Computer used for storage and dispensing of digital information. An example is a video server used to store and send video for a video-on-demand service.

set-top box. The receiver in a cable television distribution system. Also called the digital converter.

shadowing. video effect on high-contrast transitions

signal level meter (SLM). A measurement instrument capable of tuning over a cable television system's transmission bandwidth with sufficient resolution to measure individual signal characteristics, such as, amplitude, power, and modulation. Modern SLMs are capable of many analog and digital signal measurements required for maintaining cable and HFC systems.

signal generator. An electronic instrument that produces audio- or radio-frequency signals for test, measurement, or alignment purposes.

signal-to-noise ratio (SNR). The ratio, expressed in decibels, of the peak voltage of the signal of interest to the root-mean-square voltage of the noise in that signal.

SLM. signal level meters

SNR. *See* signal-to-noise ratio

SONET. synchronous optical network

. spectral regrowth. Spectral regrowth is another name for the effects of intermodulation distortion on a digital signal.

spectrum analyzer. A scanning receiver with a display that shows a plot of frequency versus amplitude of the signals being measured. Modern spectrum analyzers are often microprocessor controlled and feature powerful signal-measurement capabilities.

spur. *See* spurious

spurious. Any undesired signals such as images, harmonics, and beats.

standard definition television. NTSC or PAL television signals converted to digital format for distribution, often multiplexed onto a signal analog channel carrier with other televisions channels or data.

suckout. (1) The result of the coaxial cable's center conductor, and sometimes the entire cable, being pulled out of a connector because of contraction of the cable. (2) A sharp reduction of the amplitude of a relatively narrow group of frequencies within the cable system's overall frequency response.

symbol. In digital modulation, the symbol is the smallest piece of information transmitted.

symbol error ratio (SER). Symbol error ratio is the total number of symbols transmitted that are interpreted incorrectly divided by the total number of symbols transmitted.

symbol rate. Symbol rate is the number of symbols transmitted over the given time, such as, 10 Msymbols/second, meaning 10,000,000 symbols per second.

synchronization. loss of creates blank screen video effect

synchronous transmission. A method of sending information over a transmission line and separating discrete characters and symbols by a precise separation in time. Synchronous transmission offers higher throughput because it does not require the start-stop bits used by asynchronous methods.

system sweep. In the installation and maintenance of a cable system, system sweep is a measurement technique using a sweeping transmitter and a dedicated receiver to measure the amplitude characteristics of the system from one test point to another.

- T -

target. The cross hair symbol on a constellation diagram that represents the ideal symbol location.

TDM. *See* time division multiplexing

TDMA. *See* time domain multiple access

TDR. time domain reflectometer

temporal. in the subjective observation of video, the motion of objects in the picture

threshold of visibility. The bit error rate at which errors become visible in a digital video picture.

tilt. Automatic correction of changes in tilt, or the relative level of signals of different frequencies.

tilt compensation. The action of adjusting, manually or automatically, amplifier frequency/gain response to compensate for different cable length frequency/attenuation characteristics.

time slot. A digital signal that is formatted to allow several separate signals to share the same data stream. A time slot is the start and stop time in which each signal is allowed to transmit.

time domain reflectometer (TDR). A measurement instrument that uses a unique signal output to test a transmission medium, such as a coaxial cable, for impairments.

time division multiplexing. A technique for transmitting a number of separate data, voice and/or video signals simultaneously over one communications medium by quickly interleaving a piece of each signal one after another.

time domain. A view of a signal's behavior as a function of time.

time domain multiple access (TDMA). Time division multiple access is a multiplexing technique which allows several independent signals to share the same data stream, by assigning each signal to a specific time slot.

third harmonic. In a complex wave, a signal component whose frequency is three times the fundamental, or original, frequency.

TLA. Three letter acronym are shortcuts to language that help experts keep their priesthood. But in fairness, acronyms of any length help keep efficient communications among the people working on similar technologies.

transmission. Transmission is the use of an electromagnetic medium, such as coaxial cable or optical fiber, to send energy from one location to another usually encoded with information.

transport. In a digital system, transport is the transmission of program inputs to a head end.

triple beat, composite triple beat (CTB). Odd-order distortion products created by three signals, mixing or beating against each other, whose frequencies fall at the algebraic sums of the original frequencies. Synonymous with third-order beat.

- V -

VBW. *See* video bandwidth

vector. A vector is line of specified length and pointing is a specified direction that represents a signal's magnitude and phase. Usually vectors represent the modulation of a carrier signal.

vector diagram. A plot of a signal's magnitude and angle shown as Cartesian coordinates in terms of incidental, I, and quadrature, Q.

vectorscope. The vectorscope is used to show the vector diagram produced by a color-bar signal, and by means of a suitably marked template or cursor, laid on the face of the tube, the signal can be checked to determine whether or not it is within specified tolerances. It also used to show the effects of nonlinearities which results in modulating of the chrominance signal by the luminance signal.

vector signal analyzer. A measurement instrument which analyzes complex signals in time, frequency, and I/Q domains, usually by performing Fourier analysis on the input signal's time waveform.

vertical interval test signal (VITS). A signal that may be included during the vertical blanking interval to permit in-service testing and adjustment of video transmission.

vestigial sideband. Amplitude modulation in which the higher frequencies of the lower sideband are not transmitted. At lower baseband frequencies, the carrier envelope is the same as that for normal double-sideband. Vestigial sideband can be used to transmit analog- or digital-coded information baseband.

video. A term pertaining to the bandwidth and spectrum of the signal that results from television scanning and that is used to reproduce a picture. In spectrum analyzer operation, a term describing the output of a spectrum analyzer's envelope detector. The frequency range extends from 0 Hz to a frequency that is typically well beyond the widest resolution bandwidth available in the spectrum analyzer. However, the ultimate bandwidth of the video chain is determined by the setting of the video filter. Video is also a term describing the television signal composed of a visual and aural carriers.

visual carrier. The visual carrier is the portion of an analog television signal that contains the picture. A television signal contains both a visual and an aural carrier.

VITS. vertical interval test signal

VOD. video on demand

VSB. *See* vestigial sideband

- W -

WAN. *See* wide area network

. waveform monitor. A special-purpose oscilloscope which presents a graphic illustration of the video and sync signals, amplitude, and other information used to monitor and adjust baseband video signals.

wide area network (WAN). An integrated data network linking metropolitan or local networks over common carrier facilities.

Appendix B—Performance and Measurement Map

System performance is measured by specific parameters, such as C/N and distortion. Parameters judge performance against standards established for the system's design. System behavior is broader collection of system attributes which has less-documented parameters to judge performance by. Examples are the amount of system impairment from one test point to another and the level of ingress in the return path. The performance and measurement map attempts to connect all these parameters and behaviors. It is not inclusive, and may be confusing to read at first. Certainly there are areas that already need revision as system architectures change performance and technologies.

Parts of this map are used to illustrate the parameters and measurements made in the specific measurement-oriented chapters.

Using the digital signal performance influence map

Performance of a system is shown as a solid box. The measurement or behavior related to that performance is noted by a dashed box. In the measurement chapters, the instrumentation used for each measurement is included in the diagram, but omitted here because of the space available.

Every performance attribute of a cable television system is influenced by some, and influences others. The measure of the attribute is usually attached on the side of the performance. As an example of its use, find the "Carrier-to-noise ratio (C/N)" solid box toward the left center of the diagram. To the left, the influences, you see two lines merging: one from the analog carrier, and one from the noise. Simple. To see the performance that is influenced by C/N, look to the line exiting on the right. It goes to "Analog Channel Quality." System noise certainly affects digital signal performance, but good analog performance, such as the proof-of-performance standards in the U.S. are a prerequisite for good digital performance. (All the POP specs could not fit in this figure!)

Now look at "In-channel noise & distortion" found towards the upper right corner. Seven influences arrive from the left: "Out-of-channel distortion," "CIN," "CSO/CTB," "Interference," "Noise," and

"Group delay." Out-of-channel distortion of a signal certainly influences the in-channel performance of another. CIN, CSO/CTB, etc, are important influences, too. The result of in-channel noise & distortion is poor error rates in the digital channel.

The most important influence is shown as the line between the "Analog Channel Quality" and the "Digital Channel and Data Quality" boxes. This simply means that good analog performance is a necessary starting point for adequate digital performance.

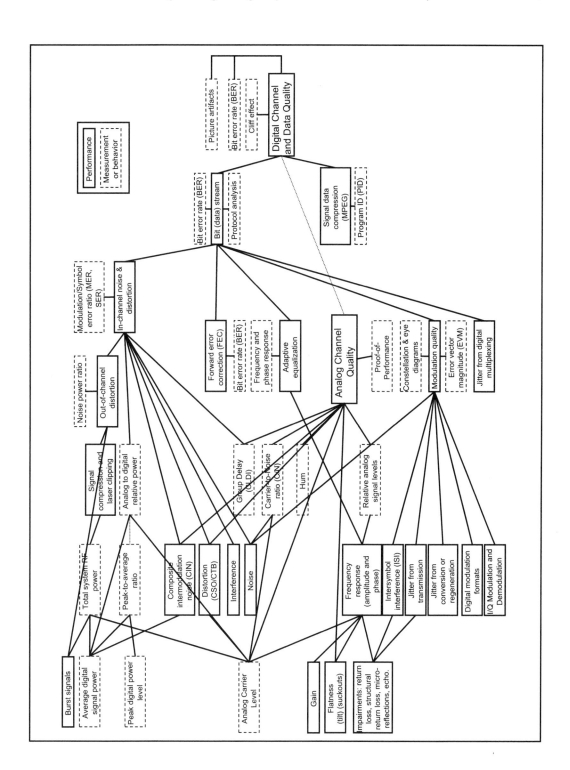

Appendix C—Equipment for Testing Digital Signals in Cable TV Systems

This appendix is an overview of the wide range of equipment for testing the fundamental parameters of digital signals in cable television distribution systems.

What you will learn

- Can you tell the primary purpose of a test instrument from its name and its input connector?
- What is the difference between a spectrum analyzer and a cable analyzer?
- Where and when are these instruments best used?
- Which instruments make digital measurements?
- What is the relative ease-of-use, accuracy, and cost of each?

Input connector identification

A test instrument's front panel label and its input connector style can tell you a lot about the type of measurements and applications it offers. A summary of these services and connector types is shown in Figure Figure C.1.. Input connectors are designed for specific types of signals and exclude others. The instrument for a digital data stream has multiple pin connectors for parallel transmission of bits, and an RF connector, LAN/WAN connector or twisted pair for serial data bits. RF connectors, such as phono, video, F, and other coaxial cables are used for everything from baseband to broadband RF. An optical connector is usually for a single laser.

Services	Connectors	Measurement Uses
Network and data streams	LAN/WAN connector 10 Mbps Multi-pin 10 Mbps	Bit rates, frame, and packet error monitoring, and protocol analysis
Audio Frequencies	Phono jack 1 MHz Video 9 MHz	Audio and video analysis, vector color analysis, and baseband video testing
Video/RF to Microwave	Type F 1000 MHz BNC 1000 MHz Type N 26 GHz	Satellite feed test, cable TV analog testing (proof-of-performance), video modulation analysis, channel and total power, constellation diagrams, modulation quality testing, and RF time domain reflectometry
Microwave to millimeter	APC-7, 3 mm 50 GHz Waveguide 300 GHz in octave bands	Antenna and antenna feedline testing
Lightwave	Fiber optic 1300, 1500 nm	Optical power testing and optical time domain reflectometry

Figure C.1. Connectors and their uses in testing.

These are the instruments for testing in data stream:

- Bit error rate tester (BERT)
- Protocol analyzer

These instruments test in the video and RF frequency ranges:

- Cable TV analyzer
- Cable sweep test system
- Digital video power analyzer
- Modulation analyzer
- Modulation domain analyzer
- Peak power meter
- Picture quality analyzer
- Power meter

- Signal level meter (SLM)
- Spectrum analyzer
- Vector signal analyzer

For testing the fiber and the power in optical links:

- Optical time domain reflectometer (OTDR)
- Optical power meter

Ease of use	R & D
Accuracy	excellent
Cost	very high

The following sections rate each type of instrument in the above lists according to their applicability to the measurements in the performance map. A rating block, like the one in this paragraph, accompanies each product summary to give you an idea of the product's ease, accuracy, and cost for the measurement parameters listed. R & D in the ease-of-use line means that the product requires expert knowledge to extract good information from the measurement.

The following sections list the products used to evaluate digital signals in cable television systems. Each product has a brief description, the parameters they measure, a typical connector used for its input, and an application suitability rating box. The test equipment is listed in alphabetical order.

BNC
connector

Bit error rate tester (BERT)

Ease of use	easy
Accuracy	very good
Cost	low

It is no surprise that bit error rate testers measure bit error rates. They are fed the data stream directly to read and analyze the error rate compared to a known code sequence. The known code sequence is either fed to the system under test by the BERT itself, or by another bit generator in the system. The bit sequence used for testing can be proprietary to the system, or one of many standard bit sequences, known as a PRBS, for pseudorandom bit sequence. The input to a BERT is either serial, as a serial data stream through a single coaxial cable, twisted pair, or parallel cable.

BNC
connector

BERTs are available with a great number of features, connectors, standards, and interface standards. Most are available with bit generators to provide the test bit streams as well as bit streams with errors intentionally inserted. More complex test equipment, such as digital video modulation analyzers, offer BERT testing as a built-in function to make more efficient use of a single box.

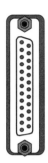

Using a BERT to test end-to-end or real-time margin for a digital signal in your system must be built into the system as part of its design or reengineering. Think through the testing needs, the test points, and purpose of BER testing before finalizing your system specifications. Once designed into a system, the BERT should serve well as a portable, easy-to-use, and low-cost addition to your maintenance test equipment.

Cable television analyzer

**Type F
connector**

Ease of use	excellent
Accuracy	very good
Cost	moderate

Good digital signal quality begins with good analog signal performance. The most versatile and efficient tool for making a host of analog signal and system tests, known as proof-of-performance testing in North America, is the cable television analyzer. The cable television analyzer is part spectrum analyzer, part TV tuner, and part television waveform analyzer, and is adapted for NTSC, PAL and SECAM measurements. The RF tests include:

- Visual and aural carrier levels and frequencies
- Depth of modulation
- Aural and FM broadcast carrier deviation
- Carrier-to-noise-ratio

Video measurements required are:

- Differential gain
- Differential phase
- Chrominance-to-luminance delay inequality

Usually the cable TV analyzer can be configured to provide automatic, unattended monitoring of the vital system and signal parameters.

Ease of use	good
Accuracy	very good
Cost	moderate

The rating box at the left applies to the cable TV analyzer, such as the Hewlett-Packard 8591C, when it is used for digital signal measurements. As Chapter 9 and Chapter 10 demonstrated, cable TV analyzers can be used to make RF measurements on digital signals, although not all of these are available as one-button tests today. The manual and semi-automatic measurements of digital signals include:

- Average power
- Adjacent channel power (ACP)

- Carrier-to-noise
- Composite intermodulation noise
- Noise power ratio (NPR)
- Peak power
- Peak-to-average power

Digital video power analyzer

Ease of use	installation
Accuracy	excellent
Cost	moderate

The digital video power analyzer is a specialized power meter, with features and performance aimed at fully characterizing the power profiles of digital signals on RF cables. Many of the analyzer's special features, such as characterization of the digital signal in the time domain and dual power inputs, are required for design, occasionally at installation, but are not usually required for routine maintenance.

Type N connector

However, most video power analyzers are used to measure average power, peak power, and peak-to-average ratios. Their power sensors collect power information just like general-purpose power meters, but can respond to the instantaneous power changes in digital signals, including peak power. Since the inputs to these analyzers are wideband, up to 40 GHz, depending upon the power sensor, the test point must be restricted to a single channel, or a tuned bandpass filter must be used to exclude other signals.

When selecting a video power analyzer for setting up the return path burst signals, make sure the analyzer can cover the full frequency range of interest. Most power meter sensor technologies may not be specified down to the 5 MHz low end of the return's frequency band.

Digital video modulation analyzer

Ease of use	very good
Accuracy	very good
Cost	moderate

The digital video modulation analyzer, such as the DVB-C QAM analyzer, such as the 8594Q from Hewlett-Packard, is a digital signal receiver that can look at the signal's modulation, modulation quality, obtain information about the quality of signal's transmission path, and its margin. It is used primarily for installation and maintenance because of its broad qualitative and quantitative measurement capability.

Type N connector

As digital signals become standard in world-wide cable systems, digital video modulation analyzers will be designed to meet the new standards with measurements tailored to the modulation formats. Here is a list of those available on the QAM analyzer:

- Average channel power
- Adjacent channel power (ACP)
- Constellation display
- Error vector magnitude (EVM)
- Equalizer response
- Data measurements (e.g., data stream content, checking proper MPEG decoding)
- I/Q offset
- Modulation error ratio (MER)
- Spectrum analyzer measurement capability (e.g., spectral views, distortion measurements)

Because the video modulation analyzer is based on a spectrum analyzer platform, it can be used over the entire RF transmission path, forward and return.

Modulation analyzer

Type N
connector

Ease of use	good
Accuracy	excellent
Cost	moderate

The analog modulation analyzer characterizes common modulation forms, such as AM, FM and PM. It combines the functions of several RF instruments, covering from below 1 MHz to over 1 GHz, to give complete, accurate channel and adjacent channel power measurements. The general purpose modulation analyzer does not make the peak power measurements necessary for digital peak-to-average evaluation.

Modulation domain analyzer

Type N
connector

Ease of use	R & D
Accuracy	very good
Cost	high

The modulation domain analyzer views signal modulation as phase, frequency, or time interval versus time, to allow a unique view of digital modulation formats that rely on timing to maintain their high quality. The analyzer is based on a counter/timer instrument platform, with built-in statistical analysis graphics for viewing results, such as jitter.

The modulation domain analyze is primarily a design tool for R&D.

Optical time domain reflectometer (OTDR)

Ease of use	very good
Accuracy	no digital
Cost	low

OTDRs for the optical links in HFC systems are plentiful and have a number of features to choose from. Besides keeping uptime at a maximum, the maintenance teams keep the optical link free from faults and impairments that can cause extreme distortion in the RF signal outputs.

The ratings show that today's OTDR measures the quality of the optical transport, and has no demodulation capability for BER and other digital signal quality or margin tests.

Fiber optic
connector

Optical power meter

Ease of use	excellent
Accuracy	very good
Cost	low

Usually in a hand-held, portable package, fiber optic power meters offer simple results, inexpensively. Often options for SLMs also offer fiber optic power measurement capability to reduce the number of text boxes in the field to a minimum.

Fiber optic
connector

Optical power meters do not offer any digital measurement capability.

Peak power analyzer

Ease of use	installation
Accuracy	excellent
Cost	moderate

Peak power capability is often a function built into RF and microwave power analyzers, as described in the "Digital video power analyzer" section above. Its use, because of the complexity of applying its capability to digital cable signals in RF transport, is usually at installation of a system.

Type N
connector

Peak power meter

Ease of use	very good
Accuracy	very good
Cost	low

The power meter directly displays the peak power of RF pulses over a 100 MHz to 18 GHz frequency range. If proper channel filtering is provided, the power meter can be used to measure the peak responses of digital signals.

Type N
connector

Peak power measurements are being built into communication test sets and digital modulation analyzers to provide measurement versatility in as few test boxes as necessary. Unless the peak power meter is specialized for digital signals its measurements are primarily applicable to impulse power, as used in the return path. However, the return path requires the statistical analysis offered by power analyzers.

Picture quality analyzer

Ease of use	R & D
Accuracy	very good
Cost	high

The first quantitative picture quality analyzers are just appearing in the market. The primary goal of the analyzer is to put numbers on the subjective opinion of how good a picture looks over time and changing content. Picture complexity, data rate, and encoding algorithm changes effects can be analyzed. But for now, this is an R&D tool for the design of components carrying and processing compressed video data streams.

Power meter

**Type N
connector**

Ease of use	excellent
Accuracy	excellent
Cost	low

The general purpose power meter is an excellent instrument for measuring the average power of a digital channel, or the total power of entire cable content. Because the power sensors are wideband RF, a tunable bandpass filter needs to limit the input power to the channel under test. A suitable tap into the RF trunk or feeder line provides a real-time power level of the entire system.

Protocol analyzer

**WAN/LAN
connector**

Ease of use	network
Accuracy	very good
Cost	low

Protocol analyzers provide detailed analysis of the interchange of information between computer network communication devices. Basic models passively monitor individual network links, usually testing from the bottom of the protocol stack upward. Many can decode traffic, measure bit error rates, and provide historical data from switches, routers, and other devices.

Protocol analyzers are useful if your system to integrates computer network communications, such as the Internet, into basic subscriber or head end communications using standards such as SONET (optical) and ATM.

Signal level meter (SLM)

Ease of use	very good
Accuracy	good
Cost	loq

Type F
connector

Signal level meters, long the workhorse field tool for maintenance in cable systems, are taking on the same versatility for digital signal measurement needs. The only caution is to make sure your system and the SLM's manufacturer agree on the types of measurements required. As discussed in Chapter 5 and Chapter 6, the wide variety of formats, transport conditions, and test point possibilities make it mandatory to understand the measurement conditions before buying SLMs in quantity for your field technicians. SLMs are sure to grow with the industry and move from just measuring digital average and burst power for specific formats, to making measurements required for viewing the current margin of a signal at a drop.

Today's SLMs are capable of measuring the digital signal power of QAM, QPSK and VSB with a typical amplitude power accuracy of ±1.5 dB, over the full forward and return frequency range of most systems.

System sweep and ingress analyzer

Ease of use	excellent
Accuracy	very good
Cost	moderate

BNC
connector

An RF system sweep is a necessary part of maintaining a system's frequency response integrity in both the forward and return paths. For forward testing, a sweep transmitter at the head end or hub sends small repetitive signal over the frequency band to be tested. Portable receivers, which can use almost any RF test point in the field, gathers the sweep data and gives the operator a picture of the system's frequency response between the transmitter and the test point. Some systems offer protection against ingress, warning when ingress is present and helping the sweep technician find the source.

Minimizing suckouts, and other frequency-amplitude anomalies greatly improves digital signal margin by reducing the amount of adaptive equalization used by the receivers.

Sweep systems must also test the return paths, especially for the transmission of burst digital signals, which are sensitive to transmission linear distortion caused by flatness problems. Badly distorted bursts arriving at the head end's receiver misshapen from group delay cause high BER.

Type F
connector

Other digital signal features include the ability to measure the average power of digital carriers, such as return path TDMA (burst) carriers.

Sweep test systems have long been capable of testing without disrupting service to the subscriber. The requirement is more important now that digital signals populate both forward and return paths. Sweep testing, can, but should not disrupt digital signal. Before investing in a cable sweep system, test it against the system's most fragile digital signals, such as those delivering data, rather than video or audio, before investing in them.

Spectrum analyzer

Ease of use	good
Accuracy	very good
Cost	moderate

The spectrum analyzer is a powerful tool for measuring and troubleshooting RF broadband cable systems. A few digital signal measurements can be made on a generic spectrum analyzer, but the trend is to use the analyzer as the platform upon which are built dedicated digital/digital video analyzers for efficient and cost-effective measurements. See "Digital video power analyzer" above, and last three chapters of this book.

Vector signal analyzer

Ease of use	R & D
Accuracy	excellent
Cost	high

The vector signal analyzer is primarily a tool for research and development labs. It is used to design and test components in digital broadcast and cable systems.

Type N
connector

Vector signal analyzers provide detailed characterization of digital modulation as applied to RF carriers, using wide receiver bandwidths which are digitally filtered to prevent the reception of adjacent channels. Bandwidths up to 10 MHz are available. Their flexibility is reflected in the number of analysis tools available:

- Constellation diagrams
- Eye diagrams
- Error magnitude
- Adaptive equalization
- Peak-to-average power
- Carrier phase noise
- Waveform capture and analysis
- Versatile demodulation capability

Appendix D—Answers to Chapter Questions

Chapter 1

1. What are the characteristics of an analog signal (for the purposes of this book)? (check all that apply)
   a) continuously varying amplitude over frequency (yes)
   b) continuously varying amplitude over time (yes)
   c) signal may be discrete amplitude values over time (Yes. There is no restriction on the nature of the time signal for an analog signal; it can represent whatever it desires. It may even pretend to be a digital signal!)
   d) the energy of the signal is spread over frequency

2. What characterizes a digital signal? (check all that apply)
   a) the amplitude of the signal is limited to a set of specific voltages (yes)
   b) the amplitude levels of the signal can easily be measured by a spectrum analyzer (No. There is no relationship to a spectrum analyzer.)
   c) energy of the signal is spread over frequency (Yes. Just like an analog signal)

3. The guidelines outlined in this chapter are always true. True or False? (False. These are guidelines, or rules-of-thumb that are generally true. There are always exceptions.)

4. Generally, digital signals hold their quality better under adverse transmission conditions than analog signals. True or False? (True. The digital signal may carry with it redundant information to be used to repair the signal if it is damaged by interference.)

5. Analog and digital signals are distributed through an analog media such as the coax cable and amplifiers of a cable television distribution system. Which of the following systems are sending digital signals? Check all that apply.
   d) Computer sending print information to a local printer (Yes. Whether serial or parallel line, the computer almost always communicates with its peripherals with digital distribution)
   e) Computer sending print information to a network printer in a remote site. (Yes. Again the print information is formatted for transport to a remote printer using different, but still digital, formats or protocols on a local area network.)
   f) Hybrid fiber-coax head end to node transport in a cable televi-

sion distribution system. (Either answer is OK. The protocols maintaining the HFC may be digital, but the main information stream to the subscribers is usually analog. The answer is ambiguous because more and more distribution systems will become combinations of analog and digital.)

g) The audio cable output of a compact disk player. (Analog. The output is analog because driving a stereo system is a CD's purpose. The manufacturers of CDs could have provided a digital output, but that would have led to the pirating of content using digital technology. That's not what the owners of the material copyrights would allow.)

h) The FAX output to a telephone line. (Yes. The FAX sends digitized information just as a modem does.)

6. It is easy to tell what program material is being sent on a digital signal. True or False? (False. Digital signals are almost always encoded by formatting that hides their program material from view along the transport path. Only the receiver dedicated to the signal's specific formats can read the convert the signal to a useful form.)

Chapter 2

1. What is the one vital test for whether a signal contains information or not? _____(Information cannot be predicted by the recipient.)

2. A carrier is
   a) a continuous wave signal with infinitely narrow bandwidth
   b) a continuous wave signal with very high frequency stability
   c) a continuous wave signal with very stable amplitude
   d) all of the above (yes)

3. A form of amplitude modulation is
   a) used for transmitting analog video chrominance
   b) used for transmitting analog video luminance
   c) changes the carrier amplitude as a function of the baseband information
   d) changes the carrier frequency as a function of the baseband information
   e) a and d
   f) b and c (Yes)
   g) a and c
   h) none of the above

4. A periodic signal waveform can be reconstructed by adding together some number of CW signals. True or False? (True)

5. A baseband waveform with sharp transitions is modulated onto a carrier without filtering or other signal processing. As the waveform transitions become sharper and/or more frequent.
   a) the modulated signal spreads over a wider bandwidth
   b) there is no difference in the signal's frequency response
   c) frequency spreading is dependent upon the type of modulation used
   d) a and c (Yes. Sharper transitions require more bandwidth to be reconstructed at the receiver. Different modulation schemes utilize the bandwidth differently, but they all demand more bandwidth for more complex baseband. Phase modulation is probably the least bandwidth efficient technique.)
   e) none of the above

Chapter 3

1. When a signal is put through an analog to digital conversion, then a digital to analog conversion, some of the analog signal's information is lost. (True. But the loss may be far less than the application's requirements, that is the receiver's ability to detect the loss.)

2. What are the important parameters to set when converting an analog signal to a digital signal? (check all that apply)
   a) sample rate (yes)
   b) detection method (yes)
   c) filtering before the ADC process (Yes. But not covered to this point in this book.)
   d) how much digital storage memory is available (yes)
   e) how fast the data can be processed (yes)
   f) near real-time processing of data is required (yes)

3. For digital representation of a signal to be accurate, the sample rate should be
   a) less than two times the highest frequency of the analog signal (No. The inverse is true.)
   a) at least as high as two times the highest frequency of the analog signal (Yes. More is better in this case.)
   a) equal to two times the highest frequency of the analog signal (Close. Two times the frequency is a minimum.)
   a) more that two times the sample period (No. Besides the fact that the period is in time units, the inverse of the sample rate in Hz would put this sample rate at half the highest signal frequency.)

4. Transmission bandwidth is a simple calculation based on the number of bits per second that have to be sent. (False. It involves

the encoding, modulation format, and the robustness of the transportation of the signal.)

5. What determines the costs of transmission? (Which is most true?)
    a) the larger the bandwidth the greater the cost (True. But it's how much time the bandwidth is used that costs.)
    b) costs are proportional to the bandwidth divided by the time which they are used (No. It is times the time used.)
    c) costs are proportional to the time, number of bits, and bandwidth used (Also true. Each parameter contributes to the value of the services to your subscriber.)
    d) bandwidth alone determines the cost of transmission, especially in the cable system you own (No. The maintenance of your system is the time factor in the equation.)

Chapter 4

1. Name the four steps to distribute an analog baseband signal as a digital channel:
    a) digitize the signal using ADC
    b) modulate the baseband digital signal onto a carrier
    c) combine with other RF signals using conversion or multiplexing
    d) distribute to the subscriber on cable or HFC

2. Layering is a technique in the design and construction of computer network and digital signal technologies. True or False? (True. The way we use the term here, layering is the computer software design architecture for the construction of digital communication technologies, including digital signal and computer network composition.)

3. Digital signal layering (select all that apply)
    a) protects a signal from transmission path problems (yes)
    b) is specified by some signal formatting standard (yes)
    c) is always part of a computer network layering standard (no)
    d) none of the above (no)

4. Match the analog/digital system building block, listed a-f, with the acronym or term g-l, that more closely belongs to its technology.
    a) Analog modulation
    b) Computer network transport
    c) Digital formatting
    d) Analog transport)
    e) Analog modulation
    f) Baseband recovery
    g) protocol (b)
    h) QAM (c)

i) X.25 (b)
j) PAL (a)
k) set-top box (f)
l) up conversion (d)

5. Multiplexing
   a) is a technique for putting more than one signal on a single electrical line (yes)
   b) is a general term for combining signals in broadband systems as well as in single channels (yes)
   c) helps conserve frequency bandwidth (yes)
   d) all of the above (yes)

Chapter 5

1. Name the three most-used digital modulation formats in North America, and their full names:
   a) QAM, quadrature amplitude modulation
   b) QPSK quadrature phase shift keying
   c) VSB vestigial sideband

2. Multiplexing is often combined with digital modulation in communications systems. (True or False?) (true)

3. Name the digital modulation format most suitable for each application:
   a) Over-the-air broadcast (VSB)
   b) Return path digital communication (QPSK)
   c) Forward path cable broadcast (QAM)
   d) Satellite direct broadcast (QPSK)

4. Which format is least susceptible to low signal-to-noise ratios?
   a) 8-QAM (Yes. The wider relative spacing of the symbols makes receiver selectivity easier in the presence of transmission impairment.)
   b) 16-QAM
   c) 32-QAM
   d) 64-QAM
   e) None, they are equally susceptible

Chapter 6

1. A vector diagram shows changes to a carrier due to modulation. If the vector is moving slowly clockwise around the origin with a constant radius: (select all that are true)
   a) there is no amplitude modulation (true)
   b) the carrier is phase modulated with a CW signal (False. A sine wave modulation of phase would have the vector waving

clockwise, then counterclockwise at the modulation frequency's rate.)

c) the frequency of the carrier is slowly increasing, but there is no
other modulation (true)

d) amplitude modulation is causing a frequency drift (False. Frequency and amplitude modulation are independent, except
when modulators fail.)

e) none of these

2. A digital signal is separated into in-phase (I) and quadrature (Q)
signals for transmission

a) to maintain high transmission quality because I and Q signals
do not interfere with one another

b) because the modulators for digital signals create I and Q signals

c) all of the above (yes)

3. A symbol is to a digital modulation signal as a letter is to an alphabet. True or False? (true)

4. A constellation diagram shows (check all that are true)

a) dots representing symbols at the symbol decision points (true)

b) dots representing symbols between decision points (False. No
symbol information can be derived from transition vectors.
Their trajectory is important in measuring peak power and
modulation efficiency, however.)

c) information that enables signal and system troubleshooting
(true)

d) all of the above

5. A digital modulation format has 16 bits per symbol. What is the
bandwidth necessary to transmit a 256 kbps digital signal?

a) 16 MHz

b) 1 MHz

c) 256 kHz

d) 16 kHz (yes, 256 kbps/16)

e) none of these

Chapter 7

1. For each of the following, tell whether the process is carried out
on the modulated signal, the digital baseband signal, or both.

a) forward error correction (baseband)

b) adaptive equalization (Baseband. Equalization is often coded
into a signal in baseband before being modulated onto the
carrier.)

c) compression (baseband)

2. What is the purpose of error correction? Choose all that apply.
   a) to correct transmission bandwidth for sharp roll-off (No. Al-
      though sharp roll-off may damage bits)
   b) to supply bits damaged in transmission (yes)
   c) correct spelling in your compliance reports
   d) to keep the signal quality as high as it can be (yes)
   e) to keep you from seeing how close to the cliff the signal is
      (yes)

3. What is the purpose of equalization? Choose all that apply.
   a) protect the signal against microreflections in the system (yes,
      or the ripple in the frequency response they cause)
   b) to protect the signal from linear distortions in the system (yes)
   c) to protect the signal from non-linear distortions in the system
      (no)
   d) to regulate taxes on your cable system

4. What is the purpose of digital compression? Choose all that apply.
   a) to save transmission bandwidth (yes)
   b) to reduce the amount of data sent while maintaining the high-
      est signal quality possible (yes)
   c) to make it possible to transmit high-quality video signals (yes)

5. The cliff effect is the tendency of a digital signal with error correction and adaptive equalization to maintain high quality even as
   impairments increase until it
   a) ceases to be transmitted (No. The cliff effect may be affected
      by transmission, but it is the receiver that fails to deliver pro-
      graming when the signal goes over the cliff.)
   b) can not be processed by the receiver (yes)
   c) slowly degrades the signal (No. The signal usually stays high
      in quality until the cliff edge is found, then the screen goes
      blank or is un-viewable.)

Chapter 8

1. A good BER indicates proper service delivery. A bad BER indicates impaired service but does not identify the cause of the
   problem. True or false? (true)

2. Which of these degrade digital signal quality? Choose all that apply.
   a) Transport impairments
   b) Distortion
   c) Interference
   d) All the above (yes)

3. Constellation and eye diagrams provide which of the following?

Choose all that apply.

a) BER (no)

b) Presence of a transport impairment (yes)

c) Baseband data stream (No. These diagrams look at a picture of the symbol stream, not the data represented by the symbols, although if the measurement equipment is set up for the specific modulation standard, the data stream could be read by a built-in demodulator.)

d) Adjacent channel power (No. Diagrams do not measure power in- or out-of-channel.)

e) A look at the quality of digital modulation, without putting a number on it. (Yes. The shape and alignment of the symbols and eye closure is a qualitative look at the signal, not a quantitative measure.)

f) A graphical display of the symbol pattern. (Yes. The constellation is the symbol pattern.)

g) A graphical display of the signal trajectory between symbols. (yes)

h) A map used by astronomers and optometrists. (no)

4. BER gives a good indication of how close a digital signal is to the cliff edge. True or False? (False, BER is an end-to-end test that only shows the signal quality as it is delivered for its end use, not how much the signal was restored by FEC and adaptive equalization.)

5. Error vector magnitude, EVM, does the following. Choose all that apply.

a) Provides a graph of the symbol position accuracy. (no)

b) Gives a percent representing how close the actual symbol points in a digital modulation signal are to the ideal symbol points. (yes)

c) Used for determining how close a signal is to the cliff edge. (Not directly. But EVM on the un-decoded signal baseband can be used to help determine the margin.)

d) None of the above. (no)

6. Modulation error ratio, MER, is used for the following. Choose all that apply.

a) A quantitative measure of how close an actual modulated signal is to an ideally modulated signal. (yes)

b) A dB power ratio of the total average symbol power to the symbol errors. (Yes. Although the language does not quote the formal definition.)

c) A value used to determine how close a signal is to the cliff edge. (yes)

d) A measure of the in-channel signal-to-noise ratio. (yes)

e) All of the above. (yes)

7. The source of signal jitter is easily determined by viewing its effect on a constellation or eye diagram. True or False? (False. The source of jitter is difficult to diagnose simply by observation of the digitally modulated signal.)

Chapter 9

1. The RMS voltage of an AC signal (true or false for each)

    a) causes the same resistor heating as a DC voltage of the same value (true)

    b) is always greater than the peak voltage (False. Never in a periodic waveform, and RMS is tough to calculate on signals that don't repeat themselves)

    c) requires a periodic voltage waveform and a constant load to calculate (true)

    d) is dependent upon the shape of the signal (False. An RMS value can be calculated for any periodic waveform, in fact, that is what makes it a valuable measurement)

    e) is always the same as average power. (False. If you consider that average power can be made for a signal, such as an analog video carrier, which varies RMS power over time.)

2. An analog amplitude modulated signal power changes as the information transmitted changes. Digital signal power remains constant over time; the information transmitted does not change the power added by the modulation. True or False? (True)

3. The power bandwidth of a digital signal (choose all that are true)

    a) determines whether the signal is staying within its channel (true)

    b) is used most often in the measurement of wireless communications (true)

    c) is usually taken for granted in cable digital signals because the signal modulators must conform to strict bandwidth standards (true)

    d) must be measured to make accurate measurements with a spectrum analyzer (true)

4. The most accurate way to measure digital channel power is

    a) with a power meter tuned to the signal center frequency (False. Generally power meters are broad band devices that cannot be tuned.)

    b) with a power meter whose input is restricted to the channel bandwidth by a bandpass filter (true)

    c) power meters are not suited to measuring digital signal power

(False. They are the best way, but most awkward to use because they need a high-quality bandpass filter.)

5. Measuring digital channel power with a spectrum analyzer can be done with (true or false for each one)
   a) the analyzer's built-in channel power function provided you adjust the analyzer's resolution and video bandwidths for the specific type of digital signal, and a correction factor. (true)
   b) the analyzer's noise marker, provided you adjust the analyzer's resolution and video bandwidths to conform to the signal's transmission bandwidth, and apply a correction factor to account for the transmission bandwidth shape. (true)
   c) the analyzer as a manually tuned receiver to collect the channel power information across its bandwidth (True, but it would be tedious beyond the value of the measurement,)

6. To measure digital channel power with a signal level meter, the SLM must have preset filters and algorithms built-in for each type of digital signal to measure. True or False? (True)

Chapter 10

1. Choose all that apply to the definition of peak power.
   a) the maximum power spikes in a digital signal (yes)
   b) the maximum envelope of the signal's amplitude (yes)
   c) the carrier level in an analog visual carrier (yes)
   d) all of the above (yes)

2. Peak power is measured by (choose all that apply)
   a) Peak power meter (yes)
   b) Modulation domain analyzer (yes)
   c) Oscilloscope (no)
   d) General purpose spectrum analyzer (Yes, with limitations)
   e) Digital video signal analyzer (yes)
   f) none of the above (no)

3. A 12 dB peak-to-average ratio is considered high. In a digital signal this high ratio suggests
   a) an efficiently operating digital modulator (Not necessarily, the ratio depends upon the type of digital modulation format. Each format has an ideal range of peak-to-average ratios.)
   b) good modulation quality (not necessarily, same answer as above)
   c) the peak power is 12 dB above the average power (True)

4. In setting the relative levels between analog video and digital video signals in a broadband cable system, use the peak power of the digital signal. True or False. (False, the average power of the

digital signal is used to compare its level with the analog video level.)

5. Burst power is used to send digital information in the return path in a cable television system. Why are burst signals used? Choose all that apply.
   a) Burst signals are less susceptible to interference. (Not necessarily, but if a data stream is broken up into different packets for multiplexing, each time and frequency packet can be used to generate a burst at a different frequency, which does improve immunity to ingress interference.)
   b) Burst signals are used in the time and frequency domain multiplexing, required to optimize the efficient use of the return path bandwidth. (True)
   c) They are the only way narrow channel bandwidths can be used. (False, burst signals can be wide.)

Chapter 11

1. What is a minimum performance requirement for protecting digital signals from distortion, noise, and interference? Choose all that apply.
   a) Conformance to analog signal proof-of-performance standards (yes)
   b) Minimized ingress and signal path corrosion (yes)
   c) High quality of the signals at the head end (Yes. Poor quality at the head end only gets worse through the system, just like an analog signal.)

2. The generally accepted power ratio between adjacent digital signal and an analog PAL or NTSC carriers is
   a) 0 to 4 dB, where the analog signal is stronger
   b) 4 to 8 dB, where the analog signal is weaker
   c) 8 to 12 dB, where the analog signal is stronger (In general, yes. A great deal of discussion is still underway about these ratios, but, because the digital signal is more robust it can survive the lower level and is less likely to interfere with the analog signal.)
   d) ingress-free and corrosion-proof signal path from head end to subscriber. (yes)
   e) none of these

3. The distortion produced by digital signals is noise-like. True or false? (True)

4. Composite intermodulation noise, CIN, is
   a) the distortion caused by mixing digital signals in the system's amplifiers (yes)

b) just like the CW characteristics of the CSO and CTB from analog channels (no)

c) likely to appear anywhere in the system's frequency range
(yes)

d) often indistinguishable from system noise (yes)

e) all of the above (no)

5. Signal-to-noise ratio for digital signals (check all that apply)

a) can always be used to predict a digital signal's margin (no)

b) can be used to predict a digital signal's margin if the system
and measurement equipment has been designed for that purpose (yes)

c) can be measured using the noise power ratio, NPR, method
(yes)

d) can be estimated using a spectrum analyzer (yes, under certain conditions)

e) all of the above (no)

6. Digital signals generally are more tolerant of system noise and
CIN than analog signals. True or False? (True, but when they fail
they crash.)

7. The best guard against interference is corrosion-proof connections and ingress-free transmission path. (True)

Appendix E—Collected Bibliography

A Guide to Picture Quality Measurements for Modern Television Systems, Tektronix, Inc., Internet address: www.tek.com/measurement/App_Note/PicQuality/picture.html, February 1998.

"CED Cable Modem Deployment Update," *Communications Engineering & Design, Chilton Publications,* New York, March 1998.

Digital Radio Theory and Measurements, an Introduction to Digital Radio Principles, Practical Problems and Measurements, Application Note 355A, publication number 5091-4777E, Hewlett-Packard Company, 1992.

"Field Installation and Maintenance Testing of DVB Systems," a paper for Hewlett-Packard Company System Engineering Training, Microwave Instruments Division, October 1996.

Interval, the SCTE Newsletter, Society of Cable Telecommunications Engineers, Inc. *DigiPoints*, Exton, PA, November/December 1996.

Interval, the SCTE Newsletter, Society of Cable Telecommunications Engineers, Inc. *DigiPoints*, Exton PA, June 1997.

Interval, the SCTE Newsletter, Society of Cable Telecommunications Engineers, Inc. *DigiPoints*, Exton PA, February 1998.

Using Error Vector Magnitude Measurements to Analyze and Troubleshoot Vector-Modulated Signals, Hewlett-Packard Company, Product note 89400-14, literature number 5965-2898E, January 1997.

Michael Adams, "The Move to HDTV, Issues to Think About," *Communications Technology* magazine, Communications Technology Publications, Inc., April 1998.

Dana Cervenka, "Designers Pour "Smarts" into Digital Test Gear," *Communications Engineering & Design*, Chilton Publications, New York, October 1996.

Helen Chen, "Testing Digital Video: A Look at Measuring Power and Interference," *Communications Technology* magazine, Phillips Business Information Inc., Rockville MD, May 1995.

Helen Chen, "Testing Digital Video: New Measures of Signal Quality," *Communications Technology* magazine, Phillips Business Information Inc., Rockville MD, June 1995.

Clyde F. Coombs Jr., *Electronic Instrument Handbook*, 2nd Edition. ISBN 0-07-012616-X. McGraw-Hill, Inc., New York, 1995.

Edwin Cooper and Neil Abramson, "A Brave New Digital World," *Communications Technology* magazine, Communications Technology Publications, Inc., October 1997.

Francis M. Edgington, "Preparing for In-service Video Measurements," *Communications Engineering & Design*, June 1994.

John Ernandez, "Understanding the Effects of Noise on Digital Signals," *Communications Engineering & Design*, Chilton Publications, New York, December 1997.

Brian Evans, *Understanding Digital TV; the Route to HDTV*, ISBN 0-7803-1082-9, IEEE Press, Piscataway, New Jersey, 1995.

Dr. Kamilo Feher and the engineers of Hewlett-Packard Ltd., *Telecommunications Measurements, Analysis, and Instrumentation*, ISBN 0-13-902404-2 025, Prentice-Hall, Inc., Englewood Cliffs, New Jersey, 1987.

Ken Freed, "QAM Delivery," *Broadcast Engineering*, Intertec Publishing, Overland Park KS, February 1997.

Roger L. Freeman,*Telecommunication Transmission Handbook*, 3rd edition, ISBN 0-471-51816-6, John Wiley and Sons, Inc.,1991.

Joe Gorin, *Spectrum Analyzer Measurements and Noise,* Application Note 1303, Hewlett-Packard Company, publication number 5966-4008E, May 1998.

William Grant, *Broadband Communications*, GWH Associates, 1996.

Ron Hranac, "DTV or Not DTV? That Is the Question," *Communications Technology* magazine, Communications Technology Publications, Inc., February 1998.

Ron Hranac, "Learn How to Make Two-Way Work," *Communications Technology* magazine, Communications Technology Publications, Inc., October 1997.

Ron Hranac, "Making Two-Way Work (Part II)," Society of Cable Telecommunication Engineers, Cable-Tech Expo '97, *Proceedings Manual: Collected Technical Papers*, Volume Two, 1997.

Louis Libin, "The 8-VSB modulation system," *Broadcast Engineering*, Intertec Publishing, Overland Park KS, December 1995.

Lawrence W. Lockwood, "VSB and QAM," *Communications Technology* magazine, Communications Technology Publications, Inc., December 1995.

M. Stephen McConnell, "Effects of analog and digital signals," *Communications Engineering & Design*, Chilton Publications, New York, December 1996.

Kenneth H. Metz, "Going Digital? Think Bit Error Rate," *Communications Technology* magazine, Communications Technology Publications, Inc., June 1997.

Daniel Minoli, *Video Dialtone Technology, Digital Video over ADSL, HFC, FTTC, & ATM*, ISBN 0-07-04-2724-0,McGraw-Hill, Inc., New York, 1995.

Jack Moran, personal communication.

Paul J. Nahin, *The Science of Radio,* ISBN 1-56396-347-7, AIP Press, Woodbury, New York, 1996.

Nicholas Negroponte, *Being Digital,* ISBN 0-679-43919-6, Random House, Inc., New York, 1993.

Harry Newton, *Newton's Telecom Dictionary*, 12th Edition, IBSN 1-57820-008-3, Flatiron Publishing, Inc., New York, February 1997.

Michael Orzessek, Peter Sommer, *ATM & MPEG-2, Integrating Digital Video into Broadband Networks*, Hewlett-Packard Professional Books, ISBN 0-13-243700-7, Prentice-Hall PTR, Inc., Simon & Schuster Company, Upper Saddle River, New Jersey, 1998.

Blake Peterson, *Spectrum Analysis Basics*, Hewlett-Packard Company, Application Note AN 150, Literature No. 5952-0292, Santa Rosa CA, 1989

Donald Raskin and Dean Stoneback, *Broadband Return System for Hybrid Fiber/Coax Cable TV Networks*, ISBN 0-13-636515-9, Prentice Hall PTR, Upper Saddle River NJ, 1998.

Dragos Ruiu, "An Overview of MPEG-2," 1996 Digital Video Test Symposium Technical Paper, Hewlett-Packard Company, 1996.

Dragos Ruiu, "The Challenges of Compressed Digital Video," 1996 Digital Video Test Symposium Technical Paper, Hewlett-Packard Company, 1996.

Deiter Scherer, "Measurement Tools for Digital Video Transmission," Hewlett-Packard Video Communications Division, *IEEE Transactions on Broadcasting*, Volume 39, No. 4, December 1993.

Boyd Shaw, *Power Measurement Basics*, 1997 Back to Basics Seminar, publication number 5965-7919E, Hewlett-Packard Company, April 1997.

David R. Smith, Digital Transmission Systems, ISBN 0-442-00917-8, 2nd Edition, Van Nostrand Reinhold, New York, 1992.

Wayne Smith, paper for the HP 89410A/41A Vector Signal Analyzer field training event, Hewlett-Packard Company, June 1996.

Oleh Sniczko, "HFC Level Measurement Procedures," TCI Communications, Inc., sent to Bill Morgan of Hewlett-Packard Company, February 27, 1997.

Andrew S. Tanenbaum, *Computer Networks*, 3rd Edition, ISBN 0-13-349945-6, Prentice-Hall, Inc., 1996.

Jeffrey L. Thomas, *Cable Television Proof-of-Performance; A Practical Guide to Cable TV Compliance Measurements Using a Spectrum Analyzer,* ISBN 3-13-306382-8, Hewlett-Packard Press, Prentice-Hall, Inc., 1995.

Matt Trezise, "Testing Digital Video Set Top Boxes and Networks," Internal Training Presentation, Hewlett-Packard, Queensferry Microwave Division, Marketing Department, February 1996

Matt Trezise, "Understanding the Measures of Signal Quality in DVB Systems," paper associated with DVB-C 8594Q application training, Hewlett-Packard Company, 1995.

Ernest Tsui and Michael Meschke, "Digital testing: QAM fundamentals," *Communications Technology* magazine, Phillips Business Information Inc., Rockville MD, May 1996.

John Watkinson, *The Art of Digital Video*, ISBN 0-240-52369 X, 2nd Edition, Focal Press, Oxford, 1995.

Robert A. Witte, *Electronic Test Instruments*, ISBN 0-13-253147-X, Hewlett-Packard Professional Books, Prentice-Hall, Inc., Englewood Cliffs, New Jersey, 1993.

Robert A. Witte, *Spectrum & Network Measurements*, ISBN 0-13-030800-5, Prentice-Hall PTR, Inc., Simon & Schuster Company, Upper Saddle River, New Jersey, 1993.

Helen Wright, *Digital Modulation in Communications Systems - an Introduction,* Application Note No. 1298, publication number 5965-7160E, Hewlett-Packard Company, 1997.

Ian Wright, *HP 8594Q QAM Analyzer Product Note, DVB-C Solutions*, Hewlett-Packard Company, Literature No. 5965-4991E, December 1996.

Index

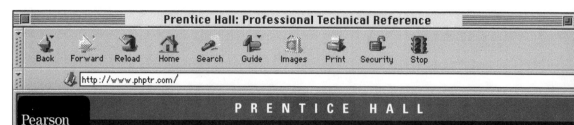

Prentice Hall: Professional Technical Reference

Back Forward Reload Home Search Guide Images Print Security Stop

http://www.phptr.com/

Pearson Education

P R E N T I C E   H A L L

Professional Technical Reference
Tomorrow's Solutions for Today's Professionals.

Keep Up-to-Date with
PH PTR Online!

We strive to stay on the cutting-edge of what's happening in professional computer science and engineering. Here's a bit of what you'll find when you stop by **www.phptr.com**:

@ **Special interest areas** offering our latest books, book series, software, features of the month, related links and other useful information to help you get the job done.

☞ **Deals, deals, deals!** Come to our promotions section for the latest bargains offered to you exclusively from our retailers.

$ **Need to find a bookstore?** Chances are, there's a bookseller near you that carries a broad selection of PTR titles. Locate a Magnet bookstore near you at www.phptr.com.

! **What's New at PH PTR?** We don't just publish books for the professional community, we're a part of it. Check out our convention schedule, join an author chat, get the latest reviews and press releases on topics of interest to you.

✉ **Subscribe Today!** **Join PH PTR's monthly email newsletter!**

Want to be kept up-to-date on your area of interest? Choose a targeted category on our website, and we'll keep you informed of the latest PH PTR products, author events, reviews and conferences in your interest area.

Visit our mailroom to subscribe today! **http://www.phptr.com/mail_lists**

www.phptr.com